Incredible Lizards

Incredible Lizards

One hundred species from around the world

Steve K Wilson

CONTENTS

ACKNOWLEDGMENTS

I owe a debt of gratitude to close friends and colleagues who have joined me on my herpetological journeys of discovery. Some travelled across the globe with me and shared the exhilaration and frustration that comes with exploring new places. Together, like babes in the wood, we entered alien forests in our search for the extraordinary. I have also been accompanied on so many trips throughout Australia, across back roads to destinations both wonderful and desolate. For sharing their time, companionship and skills, I thank: Ross Coupland, Angus Emmott, Mark Hanlon, Greg Harold, Tony and Katie Hiller, Rod Hobson, Dave Knowles, Ray Leggett, Mike Powell, Gerry Swan and Mike Swan.

I want to particularly thank Mike Swan for offering encouragement and constructive advice on the manuscript, and my wife Marilyn Parker for her ongoing support. It cannot be easy living with the perpetually obsessed!

Photo credit: Mark Hanlon

INTRODUCTION

This book is a celebration of lizard-hood. It salutes a diversity of form and function unparalleled among terrestrial vertebrates, from apex predators capable of swallowing goats to limbless subterranean insectivores. In equal measure they thrive in a tropical paradise, a waterless terrain so bleak one wonders how anything survives, and the human domain, including domestic gardens and the ruins of past civilisations. The book ventures into how lizards exploit those disparate places, and my attempts to document them around the world.

This book is born from journeys of discovery. Quests for lizards and other reptiles have taken me to some of the world's extraordinary biodiversity hot-spots in Borneo, Madagascar, Costa Rica and the Galápagos Islands. I have also crisscrossed my native Australia many times, searching for lizards in deserts, rainforests and even peoples' back yards. The maps indicate where each picture was taken.

Compiling this book has also been a voyage of rediscovery. In going over my field journals, I read notes I wrote forty years ago in a remote part of Indonesia's Kalimantan, words on pages that have rarely been turned since they were written. In another journal I relived the tragedy of loss when a bag containing all of my film, exposed and unexposed, was stolen in Madagascar.

On those earlier trips I had to nurse and guard bags containing up to 60 rolls of transparency film to be processed on my return. Of course on contemporary expeditions all the photography is digitally embedded on a chip the size of my thumbnail, with room to spare.

To select images for this book I immersed myself in those old transparencies, rediscovering images and remembering the stories behind them. It has also involved delving into my modern digital archives.

I did not choose to be passionate about reptiles. Things just happen and I often wonder why. Having an overriding obsession can be a mixed blessing but on balance, I consider myself one of the luckiest people in the world. I have been able to indulge in that intense interest and immerse myself in it as a lifestyle.

The problem was not finding 100 suitable lizards to feature, but deciding what must be left out. Those were agonising choices to make. Why 100? Because without a line under the figure I would not know how to stop!

Sadly not all of the species pictured are thriving. More than a quarter are listed on the IUCN (International Union for Conservation of Nature) Red List as being of conservation concern. But all of these 100 lizards have tales to tell.

– Steve K Wilson

HORNED LEAF CHAMELEON

Brookesia superciliaris

FAMILY:

Chamaeleonidae

LOCATION:

Analamazaotra-Perinet, Madagascar

STATUS:

Least Concern, IUCN Red List

The Malagasy people are no great fans of chameleons, but Madagascar's leaf chameleons in the genus *Brookesia* are the subjects of a particular mortal fear. It is probably based on their extremely cryptic nature.

Common features of these little lizards include lateral spines along either side of the midline, while some are adorned with horn-like processes above their eyes and many have rough or prickly skin. Against cluttered backdrops of leaf litter, with complex textures of pointed leaf tips, broken serrated edges, twig fragments and contrasting hues, the tiny slow-moving terrestrial chameleons are nearly invisible.

Traditional Malagasy ethics forbid harming animals without valid reason, and they believe spirits capable of avenging harm reside around them within the natural world. *Brookesia* are associated with 'wood-genies' having supernatural attributes. The little creatures hidden in the leaf litter are like traps set by nature for humans, akin to spiritual land mines. People fear inadvertently crushing them, and there is a popular belief that anyone who kills a *Brookesia* will suffer ill health and may die.

More than 30 species of *Brookesia* are all endemic to Madagascar. Their combined distribution includes most of the island excluding the arid south-west. However, many species are extremely range-restricted, sometimes known from single localities. The genus includes what are considered the world's smallest reptiles, with total lengths ranging from 3–11 centimetres (1–4 inches).

Unlike the colourful arboreal chameleons we generally regard as 'typical' members of the family, leaf chameleons are usually shades of brown with little ability to change colour. They also have very short tails that generally do not have the prehensile attributes we see on their more familiar relatives.

The Horned Leaf Chameleon is one of the more widespread species of *Brookesia*, occupying low- and mid-altitude rainforests along most of eastern Madagascar. Like other leaf chameleons it is terrestrial by day, but generally climbs into low shrubs to sleep at night.

As I wandered through those forests of Madagascar I could not help but wonder how many furtive little leaf chameleons I was passing unseen. Probably quite a few. I have enjoyed relatively good health in the years since my visits to the island. Perhaps this may be due, in part, to my not having accidentally harmed any *Brookesia* during my travels.

Reference: 66

MOUNT AMBER CHAMELEON

Calumma ambreensis

FAMILY:
Chamaeleonidae

LOCATION:
Montagne d'Ambre National Park,
Madagascar

STATUS:
Near Threatened, IUCN Red List

More than 40 per cent of the world's chameleon species are endemic to Madagascar. Nearly 100 species occupy virtually all habitats on an island supporting a biodiversity of continental magnitude. Tropical rainforests, deciduous forests, arid lands with succulent shrubs and spiny plants all support their own suite of chameleon species. Even urban landscapes such as the capital Antananarivo are home to chameleons.

In some areas impressive numbers of species coexist. For example, a survey of Montagne d'Ambre National Park and Forêt d'Ambre Special Reserve in the far north of the island recorded 14 species of chameleons, and some of these were undescribed. The Mount Amber Chameleon appears to be endemic to the Montagne d'Ambre National Park, where it can be found along rainforest edges.

Searching for chameleons by day can be a thankless task, even in habitats where they are known to be common. Lurking among the leaves and branches, their colours blending with the dappled light and shade, they seldom give themselves away by moving, and frequently slide themselves discreetly from view behind a branch when approached.

At night it is a different story. The beam of a torch can reveal a dozen or so in a matter of minutes, where a search of same place by day, peering fruitlessly into foliage over several hours, produced little or no signs of any chameleons.

Sleeping chameleons adopt light colours and position themselves on slender branches with their tails coiled tightly beneath them. They stand out in the torchlight and the search-image is roughly one of a pale 'snail-shape'.

In the absence of humans with spotlights, the chameleons' sleeping strategy is relatively simple. Colour is less relevant at night but location is critical. On the slender branches and foliage they are fine-tuned to the slightest disturbance, so at the first indication of any vibrations or other movements they simply let go and drop.

While protected in theory within national park boundaries, the Mount Amber Chameleon is listed as Near Threatened because of pressures from an increasing human population. These include incursions into the park to clear for agriculture such as bananas, maize and papaya, clearing trees for timber and charcoal, quarrying and zebu grazing.

Reference: 22

ANTIMENA CHAMELEON

Furcifer antimena

FAMILY:
Chamaeleonidae
LOCATION:
Belalanda, Madagascar
STATUS:
Vulnerable, IUCN Red List

The dry forests of arid south-western Madagascar are an alien, arid landscape. Nearly all of the plants grow nowhere else and many of them are formidably armed with protective spines. Appropriately called the 'spiny forests', the vegetation is dominated by succulent plants including endemic trees of the family Didiereaceae.

Yet there is something vaguely familiar about the habitat structure. In some respects those arid lands are an analogue, independently evolved on the other side of the planet, of the cactus deserts in the Americas. They share an assortment of vegetation rising from the dry soils, which use spines to guard their precious stored water.

It should come as no surprise that peculiar creatures inhabit this strange land. These include the Antimena Chameleon, which lives among the prickly shrubs and boughs in the spiny forest. When it comes to ornamentation this lizard has all the trimmings. Mature males sport a long horn on the nose, a bony casque on the head and an impressive crest of long spines along the back. And it is no lightweight, growing to a total length of nearly 40 centimetres (16 inches).

Females lack the horns and have smaller casques and dorsal crests. Very little has been documented about the Antimena Chameleon, but we know the horns on males of other chameleon species are employed for display and combat. This large mature male has scarring on its horn that is likely to have been inflicted by a rival. Males of most chameleons are notoriously intolerant of each other.

The species is restricted to a small area of Madagascar and sadly it is regarded as Vulnerable due to habitat loss and degradation. In the decade from 1990 to 2000, south-western Madagascar's dry forests suffered some of the greatest declines of vegetation cover that the island has faced. They are under pressure from rural and urban expansion and the cutting of wood for charcoal. The habitat where this lizard was photographed in the late 1980s may no longer even exist.

LABORD'S CHAMELEON

Furcifer labordi

FAMILY:
Chamaeleonidae
LOCATION:
Kirindy Forest, Madagascar
STATUS:
Vulnerable, IUCN Red List

In Madagascar's north-western deciduous forests the trees wear a thick cloak of green during the monsoon season. On the ground, masses of coloured butterflies gather at roadside puddles and biting flies drive me to distraction by attacking any exposed skin. Life is profuse and things are humming.

Labord's Chameleons are quite common in the foliage, but finding them while being assaulted by insects is quite a feat. Seldom are sex differences so stark! The female has just a stub of a horn but brilliant splashes of orange, crimson and blue (main image). The larger green-and-white male has a very high bony casque, a large prominent horn and a row of low spines along its back (inset). It is all about instant sexual recognition, for rival males to spar and pairs to mate.

It seems there is no time to lose, for they are believed to have the shortest lifespan of any terrestrial vertebrate. In the dry southern part of their range Labord's Chameleons live for just four to five months. Further north, in a more protracted and warmer wet season, they are virtually Methuselahs, reaching six to nine months. Hatchlings emerge at the onset of the annual rainy season, after spending eight or nine months as developing embryos. So they spend more time in the egg than out of it.

The young grow fast, maturing in less than two months. After mating there is a senescent decline and a near-complete population die-off. By March nearly all adults have perished, so throughout the dry season the entire population consists of developing eggs that hatch synchronously at the onset of the next rainy season.

A loss of nearly all live animals and a heavy investment in annual hatching success has evolved to exploit predicable monsoons, accompanied by a glut of small insect prey. They have placed all of their eggs in the one temporal basket, so a couple of inclement years could potentially wipe a population out. It remains to be seen whether that risky strategy continues to be effective in a changing climate, with more extreme weather patterns including protracted droughts.

Threatening processes listed for this Vulnerable species focus primarily on loss and fragmentation of its habitat in the deciduous forests of western Madagascar due to logging, conversion to charcoal and expanding agriculture. But looming over the documented threats may be the spectre of climate change.

Reference: 27

RHINOCEROS CHAMELEON

Furcifer rhinoceratus

FAMILY:
Chamaeleonidae
LOCATION:
Ampijoroa, Madagascar
STATUS:
Vulnerable, IUCN Red List

No other lizards catch their prey like chameleons. They adopt a 'slingshot' approach. Chameleons move with a slow, deliberate rocking gait, so rapid pursuit is not an option. Instead they shoot out their high-speed projectile tongues that may exceed the lizards' own body length. To successfully hit their mark they must an accurately assess direction and distance, and this is where their bizarre vision comes into play.

Chameleons are famous for their googly independently mobile eyes, bulging from their sockets and protected by scaly turrets. Those strange eyes are usually swivelling in all directions to each cover about 180 degrees, simultaneously transferring two completely disparate views of the world to the one brain. One wonders how that little brain can process such shifting, complex spatial information! But when triggered by movement, the eyes work in tandem to lock on to an insect in fine-tuned 3D vision.

Calculating its intended trajectory, the chameleon slightly opens its mouth and the club-like tongue-tip protrudes just slightly. The lizard remains motionless for what seems like an eternity, its body tensed and its attention fixed, refining direction and measuring distance. Then with mind-numbing speed, that long thin tongue is fired. The insect, adhered to the sticky tip, is hurtling back towards the chameleon's mouth and a certain fate. A chameleon's tongue measuring 14 centimetres (6 inches) can reach full extension in a sixteenth of a second so it is over in in an instant.

Rhinoceros Chameleons are appropriately named for the enormous proboscis-like horn on the nose of the males. Females lack the elaborate horn, but they can make up for this in colour. When carrying eggs they are violet with an orange tail!

This species is restricted to dry deciduous forests of north-western Madagascar. Much of that habitat has been degraded and fragmented by logging and clearing for charcoal. For this reason the Rhinoceros Chameleon is listed as Vulnerable. Fortunately it is protected within the Ankarafantsika National Park.

NAMAQUA CHAMELEON

Chamaeleo namaquensis

FAMILY:
Chamaeleonidae
LOCATION:
Welwitschia Plains, Namibia
STATUS:
Least Concern, IUCN Red List

An open gravel plain without bushes, trees or shrubs as far as the eye can see is no place for a chameleon! The only vegetation between me and the distant hills was sparse wiry little dry grass clumps and iconic *Welwitschia* with its long curling leaves like broad leathery straps. Yet there it was, looking up at me with two googly eyes.

J. E. Alexander was clearly unimpressed with Namaqua Chameleons. In 1838, when documenting his journey through south-western Africa, he wrote that they '… hissed like angry snakes, whilst a bag under their mouth swelled to great size, which, with their dark blotched bodies, gave them a hideous appearance.'

From an arboreal ancestry in lusher, more vegetated habitats, the Namaqua Chameleon has evolved to occupy deserts. One of the most arid-adapted of more than 200 chameleon species, it lives in a range of areas from stony deserts to coastal dunes nearly devoid of plants. Individuals have even been occasionally found patrolling the littoral zones below the high-tide line.

The Namaqua Chameleon has become almost entirely terrestrial, although it still retains features from those tree-climbing origins. The tail is prehensile and can be coiled but there is little for it to grasp in the desert landscape. Feet with opposable digits designed for gripping branches now walk across sand and stones. While typical chameleons often move with a slow rocking gait, Namaqua Chameleons can run quite quickly.

Namaqua Chameleons eat anything they can overpower, and they seem particularly keen on tenebrionid beetles. They also eat scorpions and lizards, and there is even a record of a chameleon killing and eating a small viper. Like other chameleons they use their high-speed projectile tongues to catch food, but they can also dash forwards and grasp prey in their mouths.

When I drove off, leaving that chameleon somewhere behind my plume of dust, I had the illogical feeling that I was abandoning a little creature in a hostile landscape. It seemed as though it needed more cover, with thicker vegetation for shelter. Yet the Namaqua Chameleon has moved beyond the realms of its relatives, and entered one of the world's harshest environments where only a select few hardy species can actually thrive.

References: 2, 85

RWENZORI BEARDED CHAMELEON

Trioceros rudis

FAMILY:
Chamaeleonidae
LOCATION:
Virunga Volcanoes, Rwanda
STATUS:
Least Concern, IUCN Red List

My guide was quite dismissive about reptiles, even informing me that there were none in the area. I found that hard to believe as we trekked through dense, richly green herbage along the flanks of Rwanda's Virunga Volcanoes. Mountain Gorillas were obviously the primary focus for the day.

So he appeared disappointed and mildly disgusted when we emerged from the forest, after a wonderful experience with gorillas, to encounter a dumpy green chameleon perched waist high in a bush by the road. He was actually repulsed when I picked it up. There are few areas across Africa where chameleons are not disliked and generally feared.

Chameleons are deeply embedded in African mythology (see also page 10). According to various beliefs they can cause death or great harm if they are touched, hiss at you or are even looked at. A Maasai friend of mine informed me that if one encounters a chameleon when embarking on a journey, one must return and start again. And despite his extensive western education, when I told him of the chameleons I had seen and handled he recoiled and asked if I acquired a rash.

The Rwenzori Bearded Chameleon is named for the mountain chain that straddles the borders of Uganda, Rwanda and the Democratic Republic of the Congo, and for the prominent row of slender spines that extend under its chin.

It is a high-altitude species, generally occurring between 2,000–4,000 metres (6,500–13,000 feet) above sea level. Much of this lofty environment has been severely degraded by slash-and-burn cultivation and cattle grazing, and the higher elevations above the frost line can experience sub-zero temperatures.

This hardy little chameleon appears to have adapted well to disturbance and harsh conditions. It remains common in roadside vegetation, and in suitable habitats where there are bushes, thickets and tall grasses it can be extremely abundant.

After photographing it I replaced it in its bush. The guide was greatly relieved, although he continued to view me as an oddity for the rest of the afternoon. There are few harmless animals as universally reviled and misunderstood as the chameleon. But to me, with their elaborate ornamentation, dazzling array of colours, feet like mittens and the strangest of eyes, they are absolute wonders of evolution.

References: 79, 85

8

SOUTHERN ROCK AGAMA

Agama atra

FAMILY:
Agamidae
LOCATION:
Augrabies Falls National Park,
South Africa
STATUS:
Least Concern, IUCN Red List

We live in a world of advertising. Billboards and other signs exhort us to buy everything from soap to life insurance. The more conspicuous they are, on well-placed elevated sites with eye-catching colour, the greater their effectiveness. Agamid lizards have been using the same strategies to promote themselves from prominent vantage points for countless millennia. The products the coloured males advertise are their sexual fitness and territorial status.

Widely known as dragons, agamids often perch with an upright stance from a raised point such as a rock, log or fencepost. Even a moderate-sized lizard, with a head-and-body length of 10–12 centimetres (4–5 inches), can be seen fairly easily from a vehicle moving at 100 kilometres (60 miles) per hour. These are the kinds of things I look out for when I am driving!

Agamids have acute eyesight, and their language is strongly based on visual cues including colour, posture and ritualised display sequences. As well as keeping a watchful eye out for each other, both sexes of many species use raised vantage points to survey their surrounds for moving prey and predators. Conspicuous perching dragons must be extremely alert, so when they sense danger they usually respond immediately and with great speed. There is normally a burrow or crevice nearby.

Males do most of the advertising, displaying eye-catching colours from prominent sites. Breeding male Southern Rock Agamas develop a distinctive pale vertebral stripe and bright blue on the face, throat and chest. They also enhance their visual appeal with sequences of head bobs and dips, but if danger threatens they flatten their heads to the substrate and the colour rapidly drains from their faces.

They are much more wary of the human form than they are of the cars we drive. I watched them through the window of my vehicle as we drove past, sitting proud on the rocks of Augrabies Falls National Park in northern South Africa. But if I left the car they would crouch or vanish. As luck would have it, a splendid male was perched near enough to the road for me to pull up, carefully wind down the window and snap some pictures with a telephoto lens.

I drove off and left him perched blue-faced and handsome. From the look of his bright colours and alert, healthy demeanour, I think he was advertising a quality product.

Reference: 15

ROUGH-TAILED ROCK AGAMA

Laudakia stellio

FAMILY:

Agamidae

LOCATION:

Parikia, Paros Island, Greece

STATUS:

Least Concern, IUCN Red List

For a Rough-tailed Rock Agama there is not much difference between a rock face, a dry-stone wall or the rough outer surface of someone's house. This arid-adapted lizard has made an easy shift into the human environment in open habitats from Egypt through the Middle East and Turkey to Greece.

More than 500 species of agamid lizards or dragons extend across the vast landmasses of Africa and Asia to Australia and the western Pacific, but the Greek population of Rough-tailed Rock Agama is at the far north-western limit of that extensive distribution. It pushes the boundary as the only species of agama that enters Europe.

A number of subspecies have been described based on physical characteristics and colouration, a testament to its variability across a fragmented range including continents and Mediterranean islands.

It is probably the most heat-tolerant species of lizard in the Mediterranean. On the hottest days, when other lizards are retreating or shuttling in and out of the shade, dragons can be seen bobbing their heads on outcrops, the crumbling walls of old fortresses, on buildings and in olive groves.

I saw them on the walls of the ancient city of Rhodes, and I watched dragons on the island of Paros, where the dominant males are liberally splashed with bright blue. To my fanciful thinking that vivid hue seemed like an acknowledgement or approval of the deep blue that features prominently in the sky, sea, painted house trimmings and roofs so typical of many Greek islands. The dragons were colour-coded to complement the scenery!

Like other dragons, they are strongly visually cued. Males court females and warn each other by bobbing their heads and inflating their bodies, and they have been observed to see and pursue insects in grass 5 metres (16 feet) away. And those blue splashes on their backs are no doubt loaded with information broadcast to any other keen-eyed dragons in their vicinity.

On rare occasions I get to combine good food with natural history. I will always remember an outdoor lunch with a sea view in the village of Parikia, dining on moussaka and drinking a dry white wine. The experience was made all the more perfect when a movement caught my eye about 30 metres (100 feet) away. Atop a stone wall was the bobbing head of a dragon.

References: 6, 86, 93

ORNATE EARLESS LIZARD

Aphaniotis ornata

FAMILY:

Agamidae

LOCATION:

Kutai National Park,
East Kalimantan, Borneo, Indonesia

STATUS:

Least Concern, IUCN Red List

At dawn in the lowland rainforests of East Kalimantan in Borneo, thin veneers of mist hug the sluggish rivers, and as the sun rises and warms the saturated air, the morning chorus begins with the whoops of gibbons and the drumming of cicadas. By midday the forest slumbers in the oppressive humidity, and after dark, with the air ringing to the new night sounds, my shirt still sticks to my back. In that perpetually warm, wet climate, sweaty clothes don't dry when you hang them up. They just rot.

Reptiles in those forests never face cool or desiccating conditions. Their body forms are less constrained by the need to soak up and conserve heat and fluids than those of their temperate relatives, so reptilian shapes are much more diverse. Snakes so slender they are nearly weightless hunt in the outer foliage and skinny lizards with long limbs like twigs cling to the vines. The Ornate Earless Lizard is an example of a tropical rainforest lizard with a morphology that could not function outside that humid, forested domain. I found just one, sleeping at night.

When I visited Kutai National Park the walking trails were rudimentary but well marked. Trekking these paths at night was a mix of joy, wonder and physical discomfort. So many exotic frogs and snakes, insects of brilliant shapes, colours and surreal sizes, all tempered by the whine of mosquitos around my head-torch and the countless bodies of so many minute, nameless creatures that drowned in the sweat of my brow and flowed with the rivulets down my face and neck.

All tribulations were instantly forgotten at the sight of a skinny little agamid lizard with a small but fancy horn on its nose asleep with limbs splayed wide across the surface of a leaf. It was so lightweight that the leaf barely registered its presence.

Little has been published about the Ornate Earless Lizard. It is endemic to the forests of Borneo, lays two eggs and the horn is present on both sexes but slightly smaller and conical on females. Of greater interest to me is the purpose of the horn. Is it for display or for camouflage to resemble a broken twig? Most records, like mine, are of lizards asleep on leaves. It will take careful daytime observations to see what Ornate Earless Lizards really get up to.

GREEN CRESTED LIZARD

Bronchocela cristatella

FAMILY:
Agamidae
LOCATION:
Sungei Buloh, Singapore
STATUS:
Least Concern, IUCN Red List

I made a big mistake with my first Green Crested Lizard. I caught it. It was among foliage beside a walking track in Sarawak, Malaysian Borneo. The gorgeous vibrant green colour drained out of it in seconds and I was left holding a drab brown lizard. I still have those old photos but it did not return to its beautiful hue while I was around. I have seen plenty more since that first encounter, and now I use a telephoto lens. There is no stress to the lizard and they keep their colour!

Green Crested Lizards are sun-loving animals that prefer edges of vegetation and disturbed areas with plenty of opportunities to bask. With extremely long and slender limbs, digits and tail, they are well designed for an arboreal life in foliage and on tree trunks. They have a vast distribution in forests, parks and gardens across South-East Asia.

For many years these were the common lizards of Singapore. They graced all the parks, home gardens and many of the street trees. But they have been edged out of some disturbed and more modified habitats by the introduced Oriental Garden Lizard (*Calotes versicolor*) (see page 32), a fast-dispersing and ecologically aggressive competitive species. Green Crested Lizards are still common in Singapore's forests and reserves, including the botanic gardens where there is plenty of lush vegetation, but they have largely lost the domestic domain.

One of the most curious aspects of this lizard is the shape of its eggs. Almost without exception, lizards' eggs are round to oval, at least when they are first laid. But Green Crested Lizards and other members of its genus lay peculiar spindle-shaped eggs, acutely pointed at each end.

There are suggestions that Green Crested Lizards have a limited ability to glide. They have been observed to jump between trees up to 7 metres (23 feet) apart with limbs and ribs splayed to reduce their fall. This is not of the same calibre as the spectacular aerial feats of true gliding dragons in the genus *Draco* (see page 34) but an impressive act regardless.

It remains to be seen how well these attractive green lizards will cope with the expanding range of the Oriental Garden Lizard. But where thick jungle grows beside paths, streams and the edges of parks they are holding on, at least for now.

Reference: 33

ORIENTAL GARDEN LIZARD

Calotes versicolor

FAMILY:

Agamidae

LOCATION:

Bandar Seri Begawan, Brunei, Borneo

STATUS:

Least Concern, IUCN Red List

I have seen them perched on the ancient carved stone of Cambodia's Angkor Wat, on the handle of a wheelbarrow propped against a hedge in Singapore, in a tangle of weedy vines in Laos, beside a market stall in Sri Lanka and atop a garden gatepost in Mauritius. More recently Oriental Garden Lizards have turned up in Brunei on the island of Borneo. These versatile reptiles have made good use of humans to create the disturbed habitats they prefer, and as an effective means of transport to expand their distribution.

Oriental Garden Lizards have followed people throughout Asia and the Indian Ocean to become one of the most abundant and conspicuous lizards within the tropical human environment across that vast range. They largely confine themselves to the areas we have impacted, where there are plenty of open spaces for basking.

Tree trunks are favoured perching sites, particularly by breeding males, which develop bright orange to red heads and forebodies with black patches on the jowls. From their prominent elevated sites they vigorously display with head-bobs and push-ups.

That red colouration of males probably lies behind the alternative name of 'bloodsucker', a term widely applied across its range. In Mauritius they are often called 'chameleons'. This caused a complete communication breakdown with my hosts, and had me rummaging through their garden looking for chameleons and complaining that I could find none. They must have seriously wondered about my ineptitude since the lizards were clearly plentiful and obvious.

Considered to be among the more invasive lizard species in the world, Oriental Garden Lizards have displaced the resident native Green Crested Lizards in Singapore (see page 30) and they may also be impacting Common Gliding Lizards (see page 34). In Mauritius they are accused of competing with the native day geckos (see page 110), and affecting some sensitive invertebrates such as phasmids. Fortunately they are reluctant to penetrate natural, undisturbed forests.

With increasing urbanisation and opening of habitats, the distribution of Oriental Garden Lizards is still growing, and this is likely to continue encroaching on various local lizard species. The males are certainly eye-catching, perched on the tree trunks and bobbing their red heads, but despite my fondness for all lizards, they are not something I am always happy to see.

References: 21, 48

COMMON GLIDING LIZARD

Draco sumatranus

FAMILY:
Agamidae
LOCATION:
Singapore Botanic Gardens
STATUS:
Least Concern, IUCN Red List

The last true flying reptile was probably a pterosaur that died around 66 million years ago. However, agamid lizards of the genus *Draco* come pretty close. I have seen them glide between trees at an inclination of nearly horizontal, swerve in mid-air to avoid saplings, and orient their bodies upwards to land vertically, head upright on a tree trunk. They can steer and adjust their momentum. Not quite flying but near enough!

The 40 species of gliding dragons occur across South-East Asia. They are arboreal, preferring well-spaced vertical tree trunks that are relatively bare of growth. Palm trees are popular. Some occupy the tallest forest trees but others thrive in parks and gardens. They grace street trees with traffic hurtling past in some of the great cities of Asia. The Common Gliding Lizard thrives in Singapore.

Gliding dragons have long hinged ribs connected to a membrane called a petagium. At rest they lie close to the slender body but when the lizard leaps from its perch to glide they are erected to support the membrane. The lizard glides swiftly with complete mastery of its trajectory, to land on another tree.

Each species has a characteristic colour and pattern combination on its gliding membrane, and a colour-coded dewlap supported by an erectable hyoid apparatus. Gliding dragons use these to communicate using a conspicuous visual language, mainly performed on vertical tree trunks where they are highly visible to each other.

There is nothing quite like the spectacle of a lizard rhythmically pushing itself in and out from the trunk, with its boldly coloured petagium expanded and its dewlap flashing a different colour. They look like odd mechanical toys! The use of species-specific colourful erectable dewlaps is mirrored by anoles in the Neotropics (see page 90) but nothing matches the hinged ribs and colourful gliding membrane.

When not gliding and displaying, the lizards are actually quite cryptic, with their dorsal greys and mottled greens closely matching the bark. They eat almost exclusively ants, which are snapped up with a quick dab of the tongue as they scurry up and down the trunks.

The Common Gliding Lizard is well adapted to the human domain. Curiously, despite its extraordinary display and gliding behaviour, and its abundance in Singapore's parks and gardens including those surrounding many high-rise housing estates, my Singaporean friends were unaware of it until I pointed it out.

JAVANESE LONG-HEADED LIZARD

Pseudocalotes tympanistriga

FAMILY:
Agamidae
LOCATION:
Cibodas, Java, Indonesia
STATUS:
Least Concern, IUCN Red List

Small green agamid lizards are common in the Cibodas Botanic Gardens and adjoining forest. Nestled on the slopes of Mount Gede in West Java at an altitude of about 1,400 metres (4,500 feet) above sea level, Cibodas is a place of beauty and tranquillity, with mild temperatures and often enshrouded in mist. It abuts submontane rainforest which is of critical importance to biodiversity on the most populated island in the world.

Each time I have visited, the conditions have been constantly cool and wet, yet the Javanese Long-headed Lizards were unconcerned. In constant drizzle, these ponderously slow-moving dragons were foraging and perching on vegetation at heights of 1–3 metres (3–10 feet) above the ground. They were well concealed on mossy trunks in the virgin forest, but they were abundant and sometimes conspicuous on trees and the strappy foliage of garden plants in the cultivated landscapes.

At night the Javanese Long-headed Lizards were easy to find in the torchlight as they sleep flat on the surfaces of leaves. It is a common strategy worldwide for small arboreal lizards to sleep on the flimsiest vegetation. Any predators that try to reach them will almost certainly disturb the foliage and allow them to drop to safety (see also pages 12 and 86).

The Javanese Long-headed Lizard occurs at altitudes of around 600–1,500 metres (2,000–5,000 feet) above sea level in central and western Java and western Sumatra. The genus, with more than 20 species, has a fragmented range, mainly confined to isolated highlands above 1,000 metres (3,300 feet) across Asia, including eastern India and southern China. Many species are extremely cryptic so they are difficult to find and are rare in collections.

Little has been written on the natural history of any species but it can be assumed they hunt invertebrates by ambush rather than active pursuit.

I was surprised to discover they never attempted to flee when I approached them, remaining motionless and just looking at me or sliding around the other side of the trunk. Excessive trust can be a dangerous thing but luckily for them I had good intentions. Tucked away in their cool, moist mountain retreat those lizards are living life in the slow lane.

Reference: 33

PHUWUA ROCK AGAMA

Mantheyus phuwuanensis

FAMILY:
Agamidae
LOCATION:
Phou Khao Khouay National Park,
Laos
STATUS:
Near Threatened, IUCN Red List

After an arduous trek in heat and high humidity through dense bamboo forests, sometimes so overgrown across the track we had to walk head-downwards through bamboo tunnels, we arrived at a slice of paradise. Clouds of multi-coloured butterflies rose and settled on the rock pavement beside a clear flowing stream tumbling over sandstone boulders against a backdrop of rainforest. The vista alone was worth the long hike. The first thing I saw before I even removed my pack was the target of my hike, a Phuwua Rock Agama.

Along the edge of the creek there are low sandstone walls with deep fissures, overhangs and caves. The Phuwua Rock Agama vanished into one of those and it was over an hour before I saw it again despite patrolling the rock faces along both sides of the creek.

Phuwua Rock Agamas have a very restricted range on the border of Laos and Thailand. Their habitat is described as rock ledges, caves and outcrops beside rainforest streams, where small groups congregate in shaded areas.

With their long slender limbs and extremely thin digits splayed wide, they scuttle over the rocks like giant spiders. They are equally sure-footed and at ease on vertical and horizontal surfaces including upside-down on the ceilings of overhangs. As I watched them fully occupying the three dimensions of their domain it was almost as though the laws of gravity did not apply. One lizard nimbly leapt from a vertical head-up posture to upside-down horizontal on an overhang ceiling. My guide told me the lizards even sleep clinging to those rock ceilings! I defy any snake, however nimble, to catch one of them off-guard at night.

It turns out that very little is known about these unusual lizards. Individuals, presumably males, have been observed to display with a small red dewlap under the throat and their ventral surfaces extended to reveal pale bluish grey. At least some of their scales are equipped with 'sensory hairs' but there are no published accounts of their purpose. They have also been noted to flatten their ribs when handled and it is proposed that this enables them to squeeze into tight crevices.

Some quality time spent with binoculars, a telephoto lens and a notepad would very likely reveal fascinating insights into this odd little lizard, which is classified in its own genus and lives in a remote paradise.

Reference: 20

MOUNTAIN HORNED AGAMA

Ceratophora stoddartii

FAMILY:
Agamidae
LOCATION:
Hakgala, Sri Lanka
STATUS:
Endangered, IUCN Red List

Horns are uncommon in the world of lizards, and when present they usually comprise some form of scaly appendage. But the spectacular horn on the snout of the aptly-named Mountain Horned Agama, or Rhino-horned Lizard, comprises just one highly modified scale. The rostral scale, positioned at the tip of most lizards' snouts, has grown into an enormous straight to curved spine. The other five members of the genus *Ceratophora* either lack horns or sprout impressive growths covered with small scales. The horns of males are largest and probably have some form of display function, perhaps among other males and to impress females.

All members of the genus are considered to be relict Sri Lankan endemics. They are restricted to the higher-rainfall areas of the south-west, which are home to many Sri Lankan endemic species. Many have extremely restricted distributions, sometimes known from single localities, and all are of conservation concern.

The Mountain Horned Agama occurs in the central hills district around 1,000–2,300 metres (3,300–7,500 feet) above sea level. Its preferred habitat is the cool and misty montane cloud forest where tree trunks sprout damp mosses. However it also extends into some disturbed areas such as well-wooded parks and gardens, cardamom plantations and even pine monocultures.

These arboreal lizards perch on vertical trunks where their green to brown colours blend well with bark, mosses and lichens. They descend to the ground infrequently, usually to feed. Studies indicate a preference for perching sites under a canopy dense enough to provide shade, and with plenty of ground cover offering abundant insect prey and protection from predators. They use the profuse mosses as refuges from temperature extremes.

Conditions were unusually hot and dry when I visited a botanical garden where Mountain Horned Agamas had been recorded as common. I eventually found a couple after some searching, and these were along the interface where the gardens abutted the cloud forest. I would like to attribute unfavourable weather to their scarcity that day but I had been told of suspected poaching from the area.

I fear the slow-moving lizards at this accessible garden site have been easy pickings. For some years it has been one of the most highly valued lizard species in the European pet trade. I suppose, like their namesakes the rhinos, these unusual-looking lizards are valued because of their horns.

References: 42, 78

LYRE-HEADED DRAGON

Lyriocephalus scutatus

FAMILY:
Agamidae
LOCATION:
Ambuluwawa Hill, Sri Lanka
STATUS:
Near Threatened, IUCN Red List

Most people who visit Ambuluwawa want to climb the impressive tower at the pinnacle of the tall conical hill and admire the vastness of the view. I went there to explore the protected forest reserve on the steep slopes of this biodiversity hot-spot near Kandy in the uplands of Sri Lanka.

Sri Lanka and neighbouring India have exchanged fauna several times via previous land bridges so it is hardly surprising that the two countries have many species in common. This particularly applies to those spread across the lowlands. But it seems the high-rainfall upland areas in the centre and south-west were subject to the longest periods of isolation by intervening dry terrain as well as by fluctuating sea levels. They harbour a suite of endemic reptile genera and species, including some of the world's most spectacular agamid lizards.

Shortly after entering the forest I became aware of a green lizard about 30 centimetres (12 inches) long clinging motionless to an upright trunk. It had a bizarre globular hump on its snout and acute bony flanges above the eyes. The slow-moving Lyre-headed Dragon belongs to a small cohort of Sri Lankan reptiles that are sometimes referred to as relict endemics. It spends most of the day perched on trunks and branches, and when disturbed it prefers to slide discreetly from view rather than dash for cover. If threatened it can distend the throat and gape the mouth to display its bright red interior.

Lyre-headed Dragons feed mainly on insects, but like many medium to large agamid lizards they augment their diet with some vegetation. Young shoots and buds have been recorded as food items. They have been observed to employ the bulbous snout to help uncover grubs and also to assist excavation of a nest hole for their clutches of three to nine eggs.

Many of Sri Lanka's relict endemics are of serious conservation concern. They have extremely limited distributions, rely on ever-diminishing rainforests, and are threatened by the usual bad culprits of habitat degradation, deforestation and encroachment by fires, with the resulting fragmentation of populations. They share an island with nearly 20 million people, or around 290 per square kilometre (750 per square mile).

The Lyre-headed Dragon occupies closed, lowland to submontane rainforests from about 25–1,600 metres (80–5,250 feet) above sea level. It has a reasonably broad distribution through the central uplands and south-west, and a degree of tolerance to habitat modification by utilising plantations and well-vegetated home gardens when these abut suitable habitat. As far as threatened species goes, it is one of the lucky ones.

References: 7, 19, 78

LAKE EYRE DRAGON

Ctenophorus maculosus

FAMILY:
Agamidae
LOCATION:
Lake Eyre South, South Australia
STATUS:
Least Concern, IUCN Red List

Some places are so mind-bogglingly hostile, it is a wonder anything survives. Consider the vast salt lakes of central-northern South Australia. Largely devoid of vegetation, they have no shade, in an arid zone where temperatures often climb above 40°C (104°F). There is no fresh water, and hot winds with dust and salt can howl unchecked across buckled crusts of salt. It takes a hardy beast to endure that blinding white nightmare. Yet the dumpy little Lake Eyre Dragon occurs nowhere else but the edges of these salinas. It is the only vertebrate to reside there permanently.

Elevated perching spots are few and far between. There are bits of driftwood deposited by previous floods, animal bones, rims of ant nests and raised lumps of salt. These are all prime real estate, generally the domain of dominant males. Subordinate males have inferior vantage points and are forced to confine their activity to the less appealing, hotter times of day.

Lake Eyre Dragons are well designed to cope in a harsh world. Their extremely pale, almost white colouration reflects heat. Their eyes are small and deeply recessed with fringed eyelids as protection from dust and moisture loss, and they are margined with black pigment to reduce glare. The tympanic membranes, exposed on most other agamids, are completely covered by scales

Beneath the salt crust a more favourable layer of fine silt provides thermally stable, humid shelter from temperature extremes. It is also a retreat when pursued, though sometimes the lizards may actually try to hide in the shadow of their pursuer!

Periodic severe rain events, sometimes unfolding more than 1,000 kilometres (600 miles) away, can dramatically affect Lake Eyre Dragons. Much of their habitat lies below sea level, so when flooding occurs within the vast Lake Eyre Basin, a catchment of nearly 1.5 million square kilometres (600,000 square miles), the water flows down and fills the lakes. It may be a decade or more between floods but their entire world is submerged, forcing whole populations onto higher ground, where they burrow into the surrounding sandy shorelines.

I have witnessed Lake Eyre in its extremes. When full it was an inland sea so vast I could not see the opposite shoreline. When empty and dry, the salt crunched under foot like giant cornflakes, a hot wind was blowing and the horizon was a shimmering haze. And there were the dragons, scampering away as I walked – unassuming little lizards thriving in one of the harshest, most unforgiving landscapes on the planet.

References: 51, 52, 95

RED-BARRED DRAGON

Ctenophorus vadnappa

FAMILY:

Agamidae

LOCATION:

Blinman Creek, Flinders Ranges, South Australia

STATUS:

Least Concern, IUCN Red List

From their elevated vantage points among steep scree slopes of fractured shales and sandstones, Red-barred Dragons scan for passing invertebrates. They also keep a wary eye on the sky for predators such as kestrels, while the males watch out for potential competitors. One observes me with what I assume to be distain as I attempt to approach it noisily and with a complete lack of composure on the loose rocks. The domain of the Red-barred Dragon is the geologically complex and scenically grand northern Flinders Ranges and associated ranges, scattered hills and outcrops.

Females and juveniles are a relatively drab reddish-brown with darker variegations. Theirs is a safe colouration to match the rocky backdrop. However a well-coloured mature male is a sight to behold, with brilliant blue along the vertebral line, black flanks with bright red bars, and throat striped blue and yellow.

Experiments have suggested that the more brightly coloured individuals are at a higher risk of predation than the duller ones. It appears that the males have evolved a dangerous trade-off, developing stunning colours that probably earn points when displaying to rivals and acquiring more mating opportunities, but at the cost of being more obvious targets for keen-eyed predators. That particularly includes raptors.

A small cohort of closely related rock-inhabiting dragon species in South Australia and adjacent New South Wales have some interesting features in common. The females and juveniles of all species resemble each other as they share colours and patterns directed towards concealment; the males have bright colours and patterns, sometimes quite stunning, that draw attention to themselves; and rival males perform ritualised, complex and highly distinctive display sequences to each other.

Opponents align themselves with bodies parallel, usually facing in opposite directions. They distend their brightly coloured throats while rhythmically raising and lowering their bodies and coiling their tails. The displays vary between species. Red-barred Dragons raise their curled tails vertically but others orient them horizontally or obliquely.

The first Red-barred Dragons I saw were captive animals when I was a kid in Melbourne. I have since visited the Flinders Ranges several times to observe them in the wild. I was always enthralled by their colours, but what is aesthetically appealing to me is critically important to the lizards. Their opportunities to mate and reproduce, and even an individual's survival, are consequences of the vibrant hues they have evolved. That bold pallette of red, blue, yellow and black means a lot more than just being pretty.

Reference: 82

YINNETHARRA ROCK DRAGON

Ctenophorus yinnietharra

FAMILY:
Agamidae

LOCATION:
Yinnetharra Station, Western Australia

STATUS:
Least Concern, IUCN Red List

In the vastness of Western Australia's interior, among stunted mulga, low shrubs and stony flats, some weathered granites look unremarkable. Rising only 1 metre (3 feet) or so above an arid landscape it would be very easy to drive past without a sideways glance.

On one of these low outcrops an agamid lizard perches upright on the heels of its feet and the claws of its fingers. It has been about 30 years since I visited this cattle station and it is good to see the Yinnetharra Rock Dragons are still here and doing well.

Approach one of these wary lizards and its launches from its rock and dashes off, its body held high on all four limbs and ringed tail well clear of the ground. Without slowing or stopping as it swerves around a bush at high speed, it vanishes into shrubs adjacent to a rock some 50 metres (160 feet) away. There could not be a more stark demonstration of a lizard's power of velocity.

Male Yinnetharra Rock Dragons are more cautious than females, with a greater critical flight distance. I cannot approach them as closely before they bolt. They are also more conspicuous. While females are dull reddish brown with darker blotches (main image), males are bluish grey with striking black and orange rings around their tails (inset). They use those as display beacons, lashing them from conspicuous vantage points.

Within that endless arid landscape there is a tiny spot, centred on low granites of Yinnetharra and Minnie Creek Stations. That is the sole known extent of the Yinnetharra Rock Dragon's range. Domed granite boulders with exfoliations and deep crevices occur to their immediate south, but Yinnetharra Rock Dragons are absent. Other agamid species are common across vast tracts of that terrain, perching on timber and rocks, tunnelling beneath them and dashing across roads. Do Yinnetharra Rock Dragons just prefer their grim landscape?

There has been little or no ecological work done, due to their remote location. Yet plenty of questions beg answers. The most basic include whether they occur anywhere else, and if not why is their range so restricted? How do they interact with each other? Are the more conspicuous males at a higher predation risk? I wondered these things when I first walked among them all those years ago, and the questions remain unanswered now. The problem with Yinnetharra Rock Dragons is that they live in a harsh landscape so far, far away!

SUPERB DRAGON

Diporiphora superba

FAMILY:

Agamidae

LOCATION:

Surveyor's Pool, Mitchell Plateau,
Western Australia

STATUS:

Least Concern, IUCN Red List

I remember camping at Manning Creek in far northern Western Australia with fellow photographer and enthusiast Gunther Schmida in the early 1980s. Browsing a recent scientific paper by eminent herpetologist, Dr G. M. Storr, I noticed that our idyllic location with its escarpments and tropical riverside vegetation was a location listed in the original description of the Superb Dragon. I picked up a stick, poked the nearest acacia bush and to my immense joy and surprise, a bright green lizard, disturbed by the shaking, slipped easily through the leaves with slow deliberate movements.

The Superb Dragon is built like a stick insect and uses the same sort of camouflage. With an extremely thin build, a long slender tail up to four times the length of the head and body, lanky long limbs like twigs and bright green colouration, it blends perfectly with foliage. This lizard is built for concealment, not speed.

Bright green, truly pea-green, is an extremely rare colour among Australian lizards. Of more than 850 species there are really only two. The range of the Emerald Monitor of northern Torres Strait Islands extends to New Guinea, so the Superb Dragon is the continent's only endemic green lizard. The scarcity of green among Australian lizards remains a mystery to me. There is no shortage of grass and foliage to hide in!

The Superb Dragon may be uniquely coloured because it is the only diurnal Australian lizard that lives almost exclusively within foliage, among the leaves and slender stems of trees and shrubs. Arboreal Australian lizards prefer trunks and branches. For some reason an extensive and promising resource has been largely ignored by modern Australian lizards. In tropical and temperate zones across the globe there are chameleons, anoles, agamids and even some geckos that exploit foliage as their primary or sole domain. Many of those are bright green.

Other members of the genus *Diporiphora* commonly venture into foliage, but only as part of a broader suite of sites including perching on rocks and stumps and dashing into thick grasses. Their protective colours are a more disruptive mix of pale stripes and dark bands to break their outlines.

I have been lucky to observe more Superb Dragons since that first memorable episode. They live in bushes, mostly growing on sandstone along the edges of creeks and gorges in Western Australia's northern Kimberley region. Their lean proportions are easily supported by the leaves and thin stems. When sleeping at night their pale bodies stand out clearly, just like sleeping chameleons (see page 13). But during the day, I rarely see them without that subtle, telltale movement of the foliage.

Reference: 51

EASTERN WATER DRAGON

Intellagama lesueurii lesueurii

FAMILY:

Agamidae

LOCATION:

Roma Street Parkland, Brisbane, Queensland, Australia

STATUS:

Least Concern, IUCN Red List

They laze in the sun, flat on their bellies beside ornamental ponds, on footpaths and even draped over the seats. Eastern Water Dragons in Brisbane's parks live easily alongside humans, paying them scant attention. But they are very interested in a team of gardeners planting flowers in the Roma Street Parkland. Each is attended by a couple of dragons, keenly watching as the spades go in, ready to snatch any worms they uncover. Further afield in a wild scenario, they would be quick to dash up a tree or leap into the water when approached.

Two subspecies of water dragons thrive in some of eastern Australia's major cities, where they can be seen near watercourses, ponds and lakes. The Gippsland Water Dragon (*Intellagama lesueurii howittii*) occurs in southern New South Wales, Canberra and eastern Victoria, and the range of the Eastern Water Dragon extends further north. Those water dragons living in Brisbane's heart are popular with tourists. They must be among most photographed local species of native animal.

The city hums with commerce. Freeways, malls and railway lines are features of inner Brisbane. These and the meandering Brisbane River separate the large water dragon populations in several established gardens as effectively as oceans around islands. Each isolated population, in its own world of curated flower beds, duck ponds and shady trees, is evolving unique characteristics.

Differences are subtle but measurable, both physically and genetically. Males in the City Botanic Gardens are the largest, while females vary in the shapes of their heads and the relative length of their limbs.

The driving forces may be their habitats, with varying amounts of low vegetation, different tree species and pond sizes. Then there is differing population density. Several hundred dragons live in the 16-hectare (40-acre) Roma Street Parkland alone. Males encounter each other more often than they would in a natural habitat so disputes including combat may be more frequent. And damaging too, as they can break each other's jaws! More males living cheek by jowl may favour larger combatants.

This is dynamic evolution in action. Of course, we are not seeing new species arising but those incremental genetic and physical shifts, measured in a few decades within a city rather than millennia between islands, illustrate just how rapidly the process can proceed.

I am drawn to Brisbane's water dragons. I look for them every time I walk through a park and there are so many tourists thrilled to snap their photos. I vote for water dragons as Brisbane's fauna emblem. These lizards, slowly changing before our eyes, are the true city slickers.

Reference: 46

BOYD'S FOREST DRAGON

Lophosaurus boydii

FAMILY:
Agamidae
LOCATION:
Lake Barrine, Queensland, Australia
STATUS:
Least Concern, IUCN Red List

It is very easy to walk past a Boyd's Forest Dragon. In north Queensland's Wet Tropics World Heritage rainforests, tourist guides keep a special eye out for this iconic species to show their guests. It features prominently in postcards, artworks and posters but it is still easily overlooked.

The lizards select vertical perches on buttresses, trunks and sapling stems, often about 1.5 metres (5 feet) from the ground. When approached they remain completely motionless or slide slowly and quietly from view.

Boyd's Forest Dragon has a close relative, the Angle-headed Dragon (*Lophosaurus spinipes*), in subtropical rainforests 1,000 kilometres (600 miles) to the south. They have laterally compressed bodies and crests of large spines. That angular, serrated outline is effective camouflage for lizards that spend extended periods motionless on vertical perches. It breaks up their shapes against the complex lines and dappled light within a rainforest setting. This follows a characteristic trend among some sedentary, arboreal lizards in rainforests around the world (see page 72).

Forest dragons are sit-and-wait predators. From their elevated perches they scan for moving prey, descend to catch invertebrates such as insects and earthworms on the forest floor, and return to their static poses on the trunks.

Most agamids are sun-loving lizards that actively seek direct sunshine to warm up, and shuttle between sun and shade to maintain an optimum operating temperature. But not the forest dragons. They are thermo-conformers. Their temperatures rise and fall in tandem with the surrounding environment. This makes sense for sedentary lizards on elevated perches in the relatively stable ambient conditions within the forest. It is not usually worth chasing errant puddles of sunshine.

Depositing their eggs is a different matter. They need sites that are warm. Historically forest dragons have buried their clutches in naturally open areas such as forest edges and clearings where trees have fallen. Now they also utilise the road edges and tracks that humans have created.

The Wet Tropics are among Australia's major tourist drawcards. While colourful butterflies and cassowaries are popular, the most iconic reptile in those rainforests is probably Boyd's Forest Dragon. It is extremely attractive and unusual, and there is a high chance that once encountered it will remain stationary while everyone takes their photos. Of course, the big question is, for every one seen, how many others slipped quietly around the other sides of the trunks?

THORNY DEVIL

Moloch horridus

FAMILY:

Agamidae

LOCATION:

Marvel Loch, Western Australia

STATUS:

Least Concern, IUCN Red List

That rotund body and upward-curved tail is instantly recognisable as a Thorny Devil crossing the track. This iconic desert oddity is always a distinctive sight for any driver as it walks with jerky movements, like a clockwork toy, across roads and tracks in Australia's arid sandy habitats.

Well protected by an armoury of large thorn-like spines, a Thorny Devil is a formidable mouthful for most predators. There is also its effective camouflage of sharply contrasting ochre blotches against a yellow to grey background, all split down the middle by a sharp white line to fragment its outline. As a last resort there is even a fake head on its neck. The bulbous hump comes complete with two large spines resembling those above its eyes. If harassed, it can tuck its head down beneath the decoy. This is unique in the world of lizards.

This slow-moving agamid has a specialist diet consisting of small black ants, primarily of the genus *Iridomyrmex*. It positions itself above an ant trail. Rank and file, as ants pass under its snout they are picked up one at time with dips of the head and dabs with the short, thick tongue. Up to 750 may be consumed in a meal.

Individuals visit their own latrine sites, easily recognisable by the scattering of faeces studded with ant remains. I have put some of those scats under the microscope. Curiously there were very few grains of sand in the mix. All those rapid dabs of the tongue and almost all a direct hit!

The Thorny Devil is a rain harvester. If rain falls on its back, or if it steps into a puddle, the effect is like blotting paper drawing water into a complex capillary system of microscopic channels between its granular scales. They direct water to its mouth allowing it to drink.

In what is hailed as an extraordinary example of convergent evolution, the Thorny Devil closely mirrors some unrelated iguanid lizards of the family Phrynosomatidae in arid North America. Some horned lizards including the Texas Horned Lizard (*Phrynosoma cornutum*) share compact spiny bodies and a specialised diet of ants. They capture them the same way along ant trails and have large stomachs to process low-nutrient prey in sizeable quantities. Most significantly they also harvest rain via capillaries between the scales (see page 78).

Evolution has taken the Thorny Devil on a crazy tangent of extreme physical and behavioural peculiarities. That is worth thinking about the next time an odd, slow-moving prickly lizard with an upturned tail appears on the road ahead.

References: 51, 95

25

GASCOYNE PEBBLE-MIMIC DRAGON

Tympanocryptis gigas

FAMILY:
Agamidae
LOCATION:
Weedarrah Station, Western Australia
STATUS:
Least Concern, IUCN Red List

The chances of seeing a Gascoyne Pebble-mimic Dragon on a bare terrain of blinding white quartz stones with sparse, stunted acacias are pretty slim. When I travelled through a remote part of arid Western Australia with fellow naturalist Mark Hanlon, our thoughts turned to this poorly known lizard as soon as we began traversing the vast tract of quartz gibbers.

Every now and then we would stop the car and I would turn random lumps of quartz in the fruitless hope of stumbling upon one. I would have had better odds pulling a needle from a haystack!

By mid-afternoon when the temperature had climbed and hope was fading I made an idle remark. 'We should be looking in the shade of those acacias.' As soon as we stepped into the first puddle of shade a small dumpy lizard, as white as the surrounding quartz, scampered at the disturbance of my footstep.

Direct mimicry of inanimate objects is rare among reptiles but some Australian earless dragons, collectively called pebble-mimics, have taken the art of camouflage to a whole new level. With round heads, plump bodies, short limbs and thin little tails they resemble the stones around them. Pebble-mimics are not particularly swift. Theirs is a strategy of 'scuttle and crouch', achieving near invisibility as soon as they stop and tuck in their limbs.

Obviously their colours match those of their respective backgrounds, but the Gascoyne Pebble-mimic Dragon has taken this a step further by copying a particular kind of rock. The little lizard is very pale brown to white in colour to specifically mimic quartz.

Pebble-mimicry is rare but not unique. It has also evolved convergently on the other side of the world. The Round-tailed Horned Lizard (*Phrynosoma modestum*), an unrelated iguanid in the family Phrynosomatidae, employs the same guise in stony deserts of the United States and Mexico (see page 80).

Gascoyne Pebble-mimic Dragons are probably quite common throughout those quartz gibbers in the mid-western interior of Western Australia. Yet very few images of them have been published. They live in a remote arid habitat and for obvious reasons they are difficult to find. We were lucky that day. On a hot afternoon, in an immense plain strewn with white stones, our ticket to success was to think like a lizard.

References: 25, 97

26

BLACK IGUANA

Ctenosaura similis

FAMILY:

Iguanidae

LOCATION:

Santa Rosa National Park, Costa Rica

STATUS:

Least Concern, IUCN Red List

The big Black Iguanas sunning themselves on logs and dry-stone walls around the Hacienda Santa Rosa are not bothered by people. Plenty of tourists visit the historic homestead and besides they are protected in Santa Rosa National Park. They look like they own the place. In a sense, they do. When the Battle of Santa Rosa was fought to oust invading Nicaraguan forces in 1856, iguanas probably witnessed the melee, and their ancestors would have been there for many centuries prior.

Black Iguanas are widespread in dry forests through western Central America between Mexico and Panama. They are abundant over much of this extensive range, despite their meat being prized in many areas, both as protein and for purported medicinal properties.

Juvenile Black Iguanas look quite different from adults. They are bright green, shifting to a more sombre hue as they grow. Their diet also changes, following a common trend among large lizard species across many families.

Carnivorous young eat mainly arthropods, growing to become omnivorous adults with a leaning towards herbivorous. The magic weights seem to lie between 50–100 grams (1.75–3.5 ounces), below which nearly all lizard species are carnivorous, and above 300 grams (10.5 ounces) where most are herbivorous. It has been suggested that the cost of catching enough prey to meet energy demands becomes more prohibitive with increasing size, so grazing accessible and locally abundant plants becomes an easier option than chasing swift and agile prey. Adult Black Iguanas eat flowers and fruits and will often climb into mango trees to take flowers directly from the plant. They augment this by catching invertebrates and small vertebrates.

Male Black Iguanas are territorial, with boundaries largely dictated by the location of females' burrows and basking sites. They have a dominance-based hierarchy, with larger and older males having access to several females on high-quality turf while sub-dominants on the periphery look for mating opportunities from females moving between burrows and feeding areas.

Burrows used for shelter are typically 1–2 metres (3–6 feet) long with room to turn around in the twisting main tunnel or its side branches. Nesting burrows are much more extensive. They may follow a complex sinuous course for up to 200 metres (650 feet) and be utilised by several females. Each deposits her eggs in a separate incubation chamber.

The Black Iguanas of Santa Rosa are lucky. If they were living outside the national park, in countries to the north and south of Costa Rica, they might well end up in the cooking pot.

References: 44, 62, 63, 71

27

MARINE IGUANA

Amblyrhynchus cristatus

FAMILY:
Iguanidae
LOCATION:
Santiago Island, Galápagos, Ecuador
STATUS:
Vulnerable, IUCN Red List

Under a tropical sun, hundreds of lizards lie basking together. Chins rest on backs. Limbs drape across bodies and tails. Occasionally a jet of hypersaline fluid squirts from a nostril. Bright red Sally Lightfoot Crabs scuttle across their flanks. The black volcanic basalt is pockmarked and pitted, sculptured here and there with whirls and peaks like whipped cream. It looks like it was poured yesterday but of course it is, literally, as hard as rock.

At more than 1 metre (3 feet) long with flat snouts, pointed conical head scales and crests of tall spines running from neck to tail-tip, there is nothing in this world remotely like Marine Iguanas. These are the world's only truly marine lizards, confined to the coastlines of Ecuador's Galápagos Islands. Thanks to their specialised diet of marine algae, they are obliged to enter the sea to feed. The species of green and red algae eaten varies between islands and seasonal abundance but about 10 genera are consumed.

Despite the Galápagos Islands straddling the Equator, the seas welling up from the southern Humboldt Current are surprisingly cold, averaging 11–23°C (52–73°F). This limits how long the lizards can actually spend in the water. They must emerge while they are still able.

Thanks to their greater mass, large males can dive deeper and remain submerged for longer. Dives up to 30 metres (100 feet) deep lasting over an hour have been recorded but most dives are shallower than 5 metres (16 feet) and briefer. At low tides, when edible algae are exposed, the iguanas can feed without needing to dive.

Extreme peaks and troughs in food availability occasionally inflict catastrophic population crashes. In response the iguanas have evolved a unique way to limit energy needs. They can shrink! During El Niño cycles when algae are in short supply, the lizards were found to have actually shrunk by as much as 20 per cent. It is believed that the very bones themselves had shortened through absorption of bone material. When food was available again they grew.

And that salty fluid squirting from their nostrils? Enlarged nasal cavities hold salt glands used to expel excess salt consumed with the algae. Expelled salt often settles on their snouts to form a white crust.

Those dozing iguanas are finely balancing critical elements of thermal management demanded by the rigours of diving into cold seas and maintaining the warmth required to digest their food. How the algae responds to ongoing climate change will be critical to their continued existence.

Reference: 94

28

GALÁPAGOS LAND IGUANA

Conolophus subcristatus

FAMILY:
Iguanidae
LOCATION:
Santa Cruz Island, Galápagos, Ecuador
STATUS:
Vulnerable, IUCN Red List

When Charles Darwin visited the Galápagos Islands in 1835 during the famous voyage of *The Beagle* he commented on the land iguanas of Santiago Island. They were so abundant, the great naturalist noted, '…we could not for some time find a spot free from their burrows on which to pitch our single tent'. Those would have been substantial burrows too, since the robust lizards grow more than 1 metre (3 feet) long.

Sometime later, following the introduction of feral pigs, goats, rats, cats, and dogs, the land iguanas of Santiago Island were completely wiped out. It is a salutary lesson that abundance is no guaranteed exemption from severe declines and extinctions.

As I walked the trails of Santa Cruz Island the iguanas browsed foliage and two males engaged in combat. They largely ignored me, not because they are tame and habituated to humans, but because they evolved in predator-free environments. To its detriment, much of the Galápagos's fauna does not know the meaning of the word 'danger' and has lost the ability to respond accordingly.

Galápagos Land Iguanas are primarily vegetarians. Prickly pear cactus (*Opuntia*) makes up about 80 per cent of their diet. They feed mainly on the fallen pads and it has been proposed that they have substantially influenced island vegetation structures.

By feeding on fallen fleshy parts they greatly reduce the asexual reproduction of the cacti, meaning fewer new plants grow from fallen bits. Yet by consuming dropped fruits they enhance the plants' sexual reproduction through seed dispersal. On islands without iguanas, cacti often grow in clumps because of asexual growth. But where iguanas are present, cactus derived from dispersed fertilised seeds are more scattered in occurrence. In effect the iguanas, along with tortoises, the other large herbivores present, have been engineering the structure of Galápagos island ecosystems.

This is the most widespread of the land iguanas on the Galápagos, yet it has suffered catastrophic declines. The lizards owe their continued existence to concerted conservation efforts including feral pest eradication. Those on Baltra Island are a re-established population, with the original inhabitants apparently wiped out in the mid-1950s by soldiers stationed there who shot them for recreation.

Between 2019 and 2021, following research by the Galápagos Land Iguana Project, more than 2,000 animals were reintroduced to Santiago Island. They were collected on North Seymour Island where an overpopulation of around 5,000 lizards is more than the local ecology can endure. I hope they thrive, and perhaps like Mr Darwin, a future camper there will have trouble finding a tent site because of all the iguana burrows!

Reference: 84

29

CENTRAL FIJIAN BANDED IGUANA

Brachylophus bulabula

FAMILY:
Iguanidae
LOCATION:
Ovalau Island, Fiji
STATUS:
Endangered, IUCN Red List

More than 8,000 kilometres (5,000 miles) of ocean is a long way between iguanas. That is the minimum distance separating the world's most isolated iguanas in Fiji from their nearest relatives in Central and South America. Just how they got there has been the subject of debate and shifting opinion.

It was originally assumed their ancestors rafted from South America on floating mats of vegetation. Iguanas have used similar means to reach the Galápagos (see pages 62, 64 and 74) but getting to the islands of Fiji is another matter. Could iguanas survive a pan-Pacific ocean journey, estimated to take six months or more, then step ashore at the other end and begin feeding and breeding?

More recently it has been proposed that they walked there when the islands were still a part of the ancient southern supercontinent, Gondwanaland. Molecular data now estimates a divergence time from their nearest relatives of 50–60 million years, old enough to place them in the area when the islands were still connected to larger landmasses. Another isolated group of iguanids, the family Opluridae in Madagascar, is also suspected to have traversed landmasses that no longer exist (see page 68).

Today there are four species of iguanas in Fiji and Tonga. They are herbivorous, eating leaves, fruit, and flowers from a range of forest plants. All are primarily arboreal and tend to occupy the highest levels of the tree canopy. The Central Fijian Banded Iguana represents this group of spectacular green lizards on moist forested islands in central Fiji.

Sadly all species are listed as Endangered to Critically Endangered. They face a multitude of challenges including ongoing habitat loss and degradation and introduced mongooses, rats and cats, and even exotic Common Iguanas (*Iguana iguana*). Those lizards are known to have detrimental effects wherever they are introduced and are now established on some islands. As well as being extremely competitive, they also confound conservation and education programs, confusing some local people about which lizards are the 'goodies' and which are the 'baddies'. Fijian iguanas also suffer from illegal collecting, turning up in tourism resorts and the international wildlife trade.

Whether their origins are as mariners rafting across ancient seas, or overland travellers on a continent now broken by the passage of time, the iguanas of Fiji now find themselves on a precipice, at the very brink of existence.

Reference: 55

30

CUVIER'S MADAGASCAR SWIFT

Oplurus cuvieri

FAMILY:

Opluridae

LOCATION:

Ampijoroa, Madagascar

STATUS:

Least Concern, IUCN Red List

The presence of iguanas in Madagascar has been furrowing brows for decades. Their closest relatives are all in the Americas, with some Pacific Ocean outliers in Fiji and Tonga (see page 66), so how did they come to populate an island off the coast of Africa? It is all the more perplexing considering the numerous species of dragons of the family Agamidae inhabiting Africa and Asia but not Madagascar. The two groups are extremely similar, exhibit countless extraordinary examples of convergence in appearance and behaviour, and have mutually exclusive distributions. They do not occur together anywhere.

Some theories place iguanas on an ancestral Africa. When Madagascar wrenched from the mainland and drifted as a bountiful ark on its own evolutionary destiny, it took some iguanas with it. Another attributes their presence on the part of Gondwana that included Madagascar when it broke up, followed by a subsequent extinction of iguanas in what is now Africa.

Evidence now suggests long-distance dispersal over water from South America, potentially via Africa or a warmer Antarctica! The evidence includes strong genetic support for a sister-relationship with the South American iguanid family Leiosauridae. The two families are believed to have diverged in the late Cretaceous to mid-Paleogene period, when Madagascar was already an island. Perhaps the ancestral rafting seafarers travelled via islands that are now long-gone.

Today there are eight iguana species in Madagascar, all in the family Opluridae. They occupy the drier habitats of the south and west, including arid spiny forests and rock outcrops. They are fast, alert lizards with an upright stance and a keen eye.

Cuvier's Madagascar Swifts are arboreal inhabitants of deciduous forests. They are common and relatively confiding, and lizards can be easily observed perching on tree trunks, stumps and logs over much of their range. From those elevated sites they adopt a 'sit and wait' approach, scanning for insect prey moving on the ground below. Individuals are typically faithful to their own favoured sites where they perch on a daily basis.

These lizards have been found to alter their responses to predators depending on what threat they are facing. They almost invariably flee from raptors but sometimes employ bluff or intimidation when confronted by snakes, lowering their dewlaps and performing push up displays.

This is one of few lizard species occurring on Madagascar that is not endemic. A separate population, described as a different subspecies, lives on Grande Comore island.

References: 41, 92

31

PLUMED BASILISK

Basiliscus plumifrons

FAMILY:
Corytophanidae
LOCATION:
Tortuguero, Costa Rica
STATUS:
Least Concern, IUCN Red List

'Flamboyant' was the word that came to mind when I first encountered a male Plumed Basilisk. In a tropical garden edged by riparian rainforest along the Rio Blanco in Costa Rica there was an unusual bird feeder. A raised horizontal plank featured several large spikes on which the owners impaled pawpaws and other fruit delicacies to attract a host of gorgeous local birds.

It also attracted a brilliantly coloured, elaborately ornamented lizard which leapt startled into mid-air and vanished into the forest as I stumbled across the bird feeder. The vista lasted only a couple of seconds but its emerald green colour and high sail-like crests left an indelible image on my mind. I later learned that it also has piercing yellow eyes! The extravagant crests are a feature of males. Females and young juveniles have more simple pointed crests on their heads.

Plumed Basilisks occur along Central America's Atlantic coast and hinterland and parts of the Pacific coast between Honduras and Panama. They are common near lowland waterways of Costa Rica, where I saw them basking on logs and branches along riverbanks, usually close to dense vegetation and often over water. Several attempts to get near them via land were unsuccessful. One unfortunate crack of a twig and they disappeared into the bushes or simply dropped into the river. But approaching them in a canoe was a different matter. By allowing the current to carry me slowly downstream I could drift quite close.

Basilisks are arboreal and semi-aquatic and they are excellent swimmers. They run with a bipedal gait, holding the body erect, forelimbs hanging and hindlimbs pedalling. But they are famous for another means of locomotion. They can run across water and for this reason the various species of basilisks are sometimes called 'Jesus Christ lizards'.

Specialised scaly fringes along the sides of each toe increase the surface area and assist a lizard sprinting on its hindlimbs to actually dash across the surface of water without sinking, at least for short distances. The lighter juveniles are the most adept. They can evade a terrestrial threat while lowering the risk of encountering something nasty under the water.

The four described species of basilisks are omnivorous, taking a wide variety of invertebrates, primarily insects but also other arthropods including crabs and shrimps. There is even a record of a Plumed Basilisk eating a bat. And they are fond of fruit, as evidenced by my first, most memorable sighting on the bird feeder.

Reference: 44

32

BROAD-HEADED WOOD LIZARD

Enyalioides laticeps

FAMILY:

Hoplocercidae

LOCATION:

Rio Napo, Amazon Basin, Ecuador

STATUS:

Least Concern, IUCN Red List

Walking along a narrow muddy track through lowland rainforest in the Amazon Basin I know I am overlooking countless numbers of unseen creatures. They are up trees, under leaves and logs, in rotting stumps and in burrows, protected from the elements and prying eyes. But some things are actually hiding in plain sight.

Scanning the trunks and vines I notice an irregularity on a sapling. A bulge on a stem. I see it is green with a spiny nape and it has been quietly observing my approach. A Broad-headed Wood Lizard is hugging the trunk about 1.5 metres (5 feet) above the ground, its chin pressed against the surface. As an Australian naturalist on my first visit to the Amazon I anticipate surprises, but my real amazement was its unexpected familiarity.

It was a case of biological déjà vu! The angular brow, spiny nape, laterally compressed body and the total length of about 30 centimetres (12 inches) resemble lizards I have seen before. Even the demeanour and setting were familiar; clinging motionless to a slender vertical perch in the dappled light of a tropical rainforest. Agamid lizards of the genus *Lophosaurus* (see page 54) look and behave in strikingly similar ways in Australia's rainforests while unrelated agamids of the genus *Gonocephalus* are analogues in South-East Asia. It seems the angular body plan, serrated along the back and compressed from side to side, is an ideal formula for a sedentary, semi-arboreal rainforest lizard. Different lizard families. Same design.

These small iguanas are common in primary lowland forests of the western Amazon Basin. Having established a 'search image' for them I found several more over the following days, a couple on the ground but mostly on similar slender stems.

Like their Australian counterparts, they spend much of their time 1–2 metres (3–6 feet) above the ground, typically on a thin trunk. There, they watch the world go by, motionless and ready to ambush invertebrates they spy moving below. Analysis of stomach contents has recorded primarily spiders, caterpillars, beetle larvae and grasshoppers, but they are probably opportunists that descend to catch anything that puts a foot wrong and is small enough to swallow.

That first lizard was only about 100 metres (300 feet) from my accommodation, so I was able to check on it regularly. For the most part it remained on the same sapling for nearly a week, changing occasionally but never straying more than a few metres from its original perch. I don't know how long they live for but I like to suppose it is still there.

Reference: 26

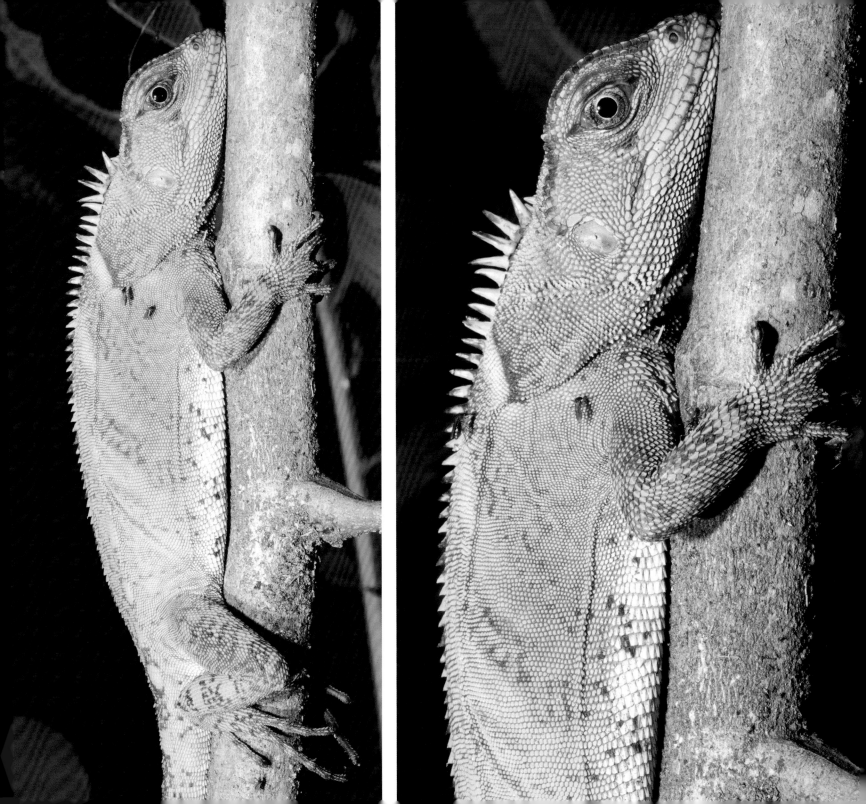

33

SAN CRISTÓBAL LAVA LIZARD

Microlophus bivittatus

FAMILY:
Tropiduridae
LOCATION:
San Cristóbal Island, Galápagos, Ecuador
STATUS:
Near Threatened, IUCN Red List

Like every other terrestrial vertebrate native to the Galápagos Islands, lava lizards owe their evolutionary history to their extraordinarily lucky ancestors, triumphing over ridiculously small odds.

The islands have never been connected to any continental landmass. Their origins are violent and fiery, rising in the ocean as molten rock punched through a weakness in the Earth's crust. They solidified and slowly drifted eastwards towards South America. Along the journey they were visited by birds and their droppings, acquired a few windblown seeds and began to develop ecosystems. Behind them, new islands blasted into existence, fuming volcanos that followed along the same trajectory.

Somewhere out in that eastern Pacific Ocean, clinging to rafts of floating debris, perhaps tree trunks washed into the sea during storms in South America, there were lizards, insects and a host of other hapless passengers. The chances of reaching landfall, specks of islands in a vast ocean, were minimal but some made it, survived and founded new populations. Rafting is the only possible way those first lucky animals could have reached the Galápagos.

There are more than 20 species of ground lizards in the genus *Microlophus,* mostly in western South America, in countries such as Peru and Chile. The animals that washed ashore on the Galápagos have since evolved into about seven to ten endemic species depending on your taxonomic outlook. They are called lava lizards and they now extend across most of the archipelago with different sibling species on most islands.

Lava lizards are among the most conspicuous vertebrates on the Galápagos, perching perky and alert with heads held high on rocks, stumps and dry-stone walls. Males and females readily defend their turf against others of the same sex, either chasing them or performing vigorous push-up displays. Males (main image) are larger with longer spiny crests but females (inset) are much more colourful, exhibiting vivid splashes of orange over the face, chin and neck. The amount and intensity varies depending on which island they are from.

Classification of lava lizards is not yet fully resolved. Some occur on single islands, but debate remains whether some others from multiple islands are widespread species or a suite of narrowly distributed island endemics. Those are important considerations given the numerous conservation issues facing the Galápagos Islands.

The San Cristóbal Lava Lizard lives on the one island where it is abundant across a range of habitats including around the buildings of the archipelago's oldest settlement, Puerto Baquerizo Moreno.

Reference: 13

34

COLLARED TREE-RUNNER

Plica plica

FAMILY:

Tropiduridae

LOCATION:

Añangu Creek, Rio Napo,
Amazon Basin, Ecuador

STATUS:

Least Concern, IUCN Red List

Collared Tree-runners spend their active hours on the smooth trunks of the largest rainforest trees. With their bodies pressed to the trunk, head downwards and long slender limbs and digits splayed wide, they look like a peculiar hybrid. Half lizard and half huntsman spider!

They have a distinctive appearance but in terms of morphology they are actually conforming to a trend repeated countless times around the world among lizards whose domain is flat vertical surfaces.

It is a simple equation spanning multiple lizard families. Exposed vertical surfaces equals dorsally flattened body, long limbs splayed out to the sides and very long thin digits. It lowers their centre of gravity and ensures a sure-footed ability to scamper swiftly over tree trunks, rock faces, walls and fences, depending on the species' preferences.

Collared Tree-runners face little or no competition for habitat space from other lizards because few if any others live on those large smooth trunks in Amazon rainforests. They also have a restricted diet that other Amazonian lizards tend to either avoid or only take in small quantities. They feed almost entirely on ants, and there is rarely a shortage of those scurrying up the trunks.

The Collared Tree-runners rarely descend from the tree trunks, often even sleeping on them. However they do come down to lay eggs. They have very small clutches of only two or three, but with an extended reproduction period, the female produces at least two clutches. It seems the benefits of small clutches with a low mass relates to their lifestyle on vertical trunks. They can't compromise that low centre of gravity by adding too much bulk.

Eggs are deposited in palm litter and inside rotting palm logs. If a nearly full-term clutch is disturbed, they all hatch spontaneously in seconds and the young dash off immediately

While I was sitting in a canoe, watching this lizard on its lofty perch and taking pictures with my telephoto lens, I thought it would be a bonus if the animal came down to the ground so I could have a closer look. I think I would have been waiting a long time!

Reference: 89

35

TEXAS HORNED LIZARD

Phrynosoma cornutum

FAMILY:

Phrynosomatidae

LOCATION:

Road Forks, New Mexico, United States

STATUS:

Least Concern, IUCN Red List

A prickly disc fringed with slender spines and a head adorned with a horned corona. That's hard to swallow! Add short legs and a ridiculously little tail and that sums up most horned lizards. For the Texas Horned Lizard, include a pattern of dark dorsal blotches along each side of the back, highlighted behind by bold pale crescents, and sliced down the middle by a crisp white line from neck to tail. At a total length of about 18 centimetres (7 inches), it is one of the largest horned lizards.

The 21 horned lizard species, distributed from Mexico, through the United States, just into Canada, are often affectionately known as 'horny toads' because of their squat postures. Indeed, the genus name, *Phrynosoma* literally means 'toad-bodied'. They are of course lizards belonging to one of several families classified under the broad banner of iguanas.

Horned lizards are well camouflaged and they are protected by their shape and thorny armour. The last, truly extraordinary resort in their line of defence is reserved for special occasions. They can squirt blood from their eyes! This is rarely performed when confronted by humans, roadrunners, snakes and leopard lizards. But dogs and dog-like species such as coyotes and foxes often trigger a response.

Before squirting, the lizard arches its back and closes its eyes. The eyelids become swollen and engorged, and suddenly fine jets of blood, the thickness of horse hairs, squirt out for distances of up to 2 metres (7 feet). Members of the dog family find it distressing, as they shake their heads with obvious distaste.

Horned lizards eat ants, mainly harvester ants. In the process they are engaged in a dietary 'arms race'. Harvester ants produce toxins as protection from predators while horned lizards evolve mechanisms to overcome those toxins to eat the ants. Blood components from Texas Horned Lizards can neutralise lethal doses of harvester ant poison administered to mice. They have developed biochemical immunity to ant toxins.

Horned lizards have been hailed as fine examples of convergent evolution, where unrelated species evolve extremely similar physical and behavioural traits. Their counterpart is the Thorny Devil, an agamid lizard from Australia (see page 56). They share compact spiny bodies, diets comprised mainly or exclusively of ants, and a unique ability to 'harvest rain'. When infrequent rains fall in their respective arid habitats, water that contacts their skin is diverted to the corners of their mouths via capillaries between the scales. Texas Horned Lizards even share with Thorny Devils a pattern of prominent blotches divided down the middle by a white line!

References: 72, 75

36

ROUND-TAILED HORNED LIZARD

Phrynosoma modestum

FAMILY:
Phrynosomatidae

LOCATION:
Road Forks, New Mexico,
United States

STATUS:
Least Concern, IUCN Red List

The view across an arid plain in New Mexico was unremarkable. It was sparsely vegetated with low shrubs and strewn with stones and pebbles across its vast expanse. As I walked, something caught my eye. Looking closely, I first saw only a jumble of scattered stones. Then I realised one of those stones was a bit odd. It was regarding me with a gimlet eye! I had stumbled across a Round-tailed Horned Lizard.

The Round-tailed Horned Lizard is one of the smallest of the horned lizards, with a head-and-body length of just 7 centimetres (3 inches). It is unusual in several ways. Most other horned lizards are pancake-shaped with a fringe of lateral spines, but it is rotund and relatively smooth.

Horned lizards usually camouflage themselves by flattening their bodies against the ground, using disruptive colours to blend against the background, and the lateral spiny fringe helps eliminate telltale shadows. The Round-tailed Horned Lizard is different. It is a pebble mimic. With legs tucked in and rounded back slightly hunched, it is a near-exact replica of a small stone. Without those lateral spines, it casts a shadow. That's what stones do. In fact shading is even enhanced by strategically placed dark suffusions along its flanks. Across their range in the south-western United States and northern Mexico they come in a range of colour forms matching shifts in substrate hues.

Mimicking pebbles makes a lot of sense. There are plenty of them, nothing eats them and they come in all shapes and colours. A predator's ability to fix on a 'search image' is reduced, meaning Round-tailed Horned lizards make occasional meals but they are not something worth investing a lot of time searching for.

In stony deserts on the other side of the world, an unrelated group of lizards have followed the same evolutionary path. Australia's pebble-mimicking agamid lizards called earless dragons (genus *Tympanocryptis*) are the same size, share a nearly identical physical appearance and crouching posture in similar arid environments (see page 58). This is a striking example of convergent evolution.

I would love put it down to skill, to my eagle eye cleverly spotting a flaw in a near perfect disguise. But the truth is, the lizard committed a cardinal error that no camouflaged creatures can afford to make. It moved.

Reference: 75

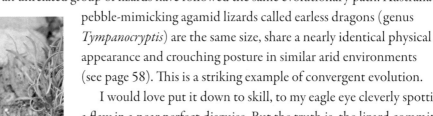

37

GREEN SPINY LIZARD

Sceloporus malachiticus

FAMILY:

Phrynosomatidae

LOCATION:

Santa Elena, Costa Rica

STATUS:

Least Concern, IUCN Red List

Along a fenceline bordering rich green dairy pastures and a patchwork of upland rainforest pockets, each fencepost had a weathered look and wore shaggy whiskers of lichens nourished by the perpetually humid air. On every fourth or fifth post perched a dumpy bright green lizard.

Green Spiny Lizards do particularly well in disturbed areas such as the edges of clearings, on stone walls and in gardens. They tend to be more common at such sites than they are in natural, uncleared habitats. In their preferred mild climates they like the open areas where they can access plenty of sunshine, at least when it is available.

Even in the balmy weather at this mid-level altitude near Monteverde in Costa Rica the lizards were warm, alert and quick to retreat into hollows when I approached them. They bask efficiently, and by using their colour and strategic postures, the lizards can maintain body temperatures up to 15°C (27°F) above the ambient air temperature.

They are common in the elevated regions of Costa Rica, from around 600–3,800 metres (2,000–12,500 feet) above sea level. At higher levels of their range young are present year-round, but below about 1,500 metres (5,000 feet) breeding is more seasonal with live young being born from late November to early March.

Males are well spaced from each other, but their small home ranges generally overlap with those of several females. They are territorial, and each is prepared to defend the central core of its home range. Much of their communication is based on colour and posture. The deep blue on the throats and chests of males is strategically placed for an easy and highly visible display to other lizards.

Speaking as an observant mammal rather than a lizard, those vibrant throats were certainly obvious to me as the Green Spiny Lizards perched erect on their fenceposts, heads held high and throats radiant in the sunshine.

References: 44, 71

38

CREVICE SPINY LIZARD

Sceloporus poinsettii

FAMILY:

Phrynosomatidae

LOCATION:

Animas area, New Mexico, United States

STATUS:

Least Concern, IUCN Red List

The rocks were black, standing out prominently with a clear junction against the surrounding geology. It was obviously an ancient larva flow. But more importantly, some of those tumbled and fissured rocks had lizards perched on them!

With fellow Australian herpetologist and long-time friend Mike Swan I was visiting this strip of basalt in western New Mexico because we had been told there were some interesting lizards living there. We had been well informed.

Heads and forebodies held high, the lizards watched us warily, and vanished quickly when approached. But they did not go far, taking refuge in the nearest crevices. Each was wedged in tight, held fast by the raised spiny scales on its body and tail. Small wonder they are called Crevice Spiny Lizards!

Crevice Spiny Lizards live only among rocks, often in sparsely vegetated areas. The population we were looking at near the Arizona border is at the far western limit of their range. From there they extend east to central Texas and south into Mexico.

Spiny lizards of the genus *Sceloporus* are extremely diverse with more than 100 species ranging from Canada to Central America, including numerous species in the United States. Thanks to their general abundance and their fondness for elevated perching sites on rocks, stumps and fenceposts, they are often among the most obvious lizards wherever they occur.

Spines and rocks are an effective mix. Rock-inhabiting lizards from many families around the world, including skinks, monitors, agamids and cordylids, have evolved tough raised spines, usually on their tails and sometimes their bodies, to wedge themselves firmly into cracks in rock and wood. In keeping with this trend, the spines of Crevice Spiny Lizards are longest and stoutest on the tail. When this is curved in front to protect the concealed body they are extremely difficult for predators to extract.

We peered into those basalt cracks and saw spiny tails blocking the crevices. I wondered how many millions of times over countless eons, lizards like those Crevice Spiny Lizards have wedged themselves in tight, daring would-be predators to try and get them out.

39

GIANT GREEN ANOLE

Anolis biporcatus

FAMILY:

Anolidae

LOCATION:

Rio Blanco, Guápiles district, Costa Rica

STATUS:

Least Concern, IUCN Red List

Being an anole is extremely dangerous. They are small to medium-sized lizards without much capacity to retaliate and are generally abundant in many habitats. This puts them on the menu for just about every carnivorous animal. For a host of snakes, mammals, birds, bigger anoles and even giant spiders, anoles are a smorgasbord waiting to be eaten, provided they can be captured.

Small wonder anoles are often cryptic and elusive. Males are obvious, sometimes briefly, when they display their coloured dewlaps (see page 90) and there are some conspicuous urban species including in the southern United States. However, for the most part they are camouflaged against their backgrounds and ready to freeze or vanish if they sense danger.

Anoles are particularly vulnerable when sleeping at night. A safety mechanism employed by nearly all anoles is to select lightweight, slender perches such as twigs and leaves as nocturnal roosts. Any slight movement that disturbs their resting place and they can drop to the ground immediately.

While spotlighting at night in Costa Rican forests, my beam traced the curvature of tree buttresses, followed the winding shapes of vines and scanned the surfaces of foliage. Giant Green Anoles stood out quite sharply in the torchlight. They were asleep on twigs and draped across the upper surfaces of leaves. I could generally find at least a few each evening.

These attractive, pea-green lizards are short-limbed and stocky compared to other anoles. With a head-and-body length of more than 10 centimetres (4 inches) they are considered relatively large. The species also has a broad distribution, extending from Mexico to northern South America.

They are strictly arboreal and occupy a range of forested habitats including disturbed areas. Vegetation along roadsides and other forest edges are popular sites where the bright green lizards bask in sunlit foliage and often perch on large understorey leaves such as those of palms, bananas and heliconias. They also stake out the leaf litter from a head-downwards position on vertical trunks, ready to drop onto passing prey.

Like all anoles, every Giant Green Anole is subject to a multitude of mortal hazards with every step it takes. But when living in tropical forests, it is a bit easier being green.

References: 4, 44, 71

40

OSA ANOLE

Anolis osa

FAMILY:
Anolidae
LOCATION:
Corcovado National Park,
Costa Rica
STATUS:
Least Concern, IUCN Red List

Something peculiar caught my eye on a tree buttress. A small brown anole was inconspicuous, and I am not sure why I noticed, but it had an unusual object in its mouth. Insectivorous lizards obviously prefer to capture prey much smaller than themselves, and they do not normally bite pieces off larger animals. So chancing upon an Osa Anole munching on the leg of a grasshopper which would have been as big as the lizard, was certainly worthy of a closer look.

Anoles are normally triggered to feed by movement, so I do wonder whether a little lizard tried its luck with a big grasshopper. For an anole with a head-and-body length of only about 5 centimetres (2 inches), the grasshopper would have been a powerful adversary with a strong kick. In the struggle, the insect lost its leg as it escaped. Or perhaps the lizard found a stray grasshopper's leg and decided to eat it without any movement stimulus to draw its attention.

Osa Anoles have a narrow distribution. They are endemic to Peninsula de Osa in south-western Costa Rica. Fortunately the richly biodiverse Corcovado National Park covers much of that area and the anoles are common in a range of habitats across the park. They live in high densities, estimated at up to 300 individuals per hectare in closed-canopy forest!

These are shrub/ground anoles that spend much of their time perched on bushes and low vertical surfaces less than 2 metres (6 feet) from the ground. The males are territorial and spend about half of their active hours displaying and fighting with each other. They reproduce throughout the year, with females laying an egg every two weeks. Courtship and mating intensify at the onset of the wet season when eggs may be deposited weekly. Osa Anoles are an annual species with very few animals surviving to a second year.

The lizard pictured may be a female. They are reported to take larger prey items, on average, than males. But the anole was stretching the normal bounds of predation and I did wonder if perhaps there was a lucky grasshopper out in that forest. One with five legs!

References: 3, 44, 71

41

AMAZON BARK ANOLE

Anolis ortonii

FAMILY:

Anolidae

LOCATION:

Rio Napo, Amazon Basin, Ecuador

STATUS:

Least Concern, IUCN Red List

The view of the Amazon rainforest canopy was breathtaking, stretched out like a vast unbroken green carpet in every direction below me. I was among the upper branches of an emergent giant, towering high above the surrounding trees. And no, I didn't haul myself up with ropes and daredevil paraphernalia. There is a lookout on this tree, accessed by a very convenient and quite easy set of steps.

On one of the surrounding boughs, festooned with lichens, bromeliads and orchids, a movement caught my eye. There was a small pale grey lizard, the same colour as the exposed bark. The Amazon Bark Anole, presumably a female, sat still long enough for me to snap a few photos before the real drama unfolded. Another anole suddenly appeared and launched into a spectacular display.

The male Amazon Bark Anole abruptly raised his body high on his thin legs, and performed rapid push-ups while extending a brilliant dewlap coloured deep orange with yellow streaks that extended well back onto the chest. A few flashes of intense colour, bobbing before the female's eyes, then the show was over and he was gone, leaving me stunned that I was able to capture the moment. The dazzling performance was over in seconds.

The anole had good reason to be brief. In a world full of predators, conspicuous displays carry obvious risks. Those rapid sequences can be regarded as flash displays. They convey enough information for their target audience, but end before a predator can get a fix on the source. Once performed, the little grey lizard is no longer obvious but the message has been delivered.

Male anoles fan their coloured throats and engage in ritualised display sequences to advertise sexual and territorial status. The dewlaps are elastic, expandable skin, normally held flat and concealed but can be erected by a moveable hyoid apparatus.

With more than 430 described species, *Anolis* is the largest genus of lizards, and to sort out who's who, many have their own colour-coded semaphores. When several distributions overlap, as is often the case, dewlap colours and patterns are critical indicators. So for anoles, and also curious humans, dewlaps are important tools to identify species.

Anoles are widespread in the Neotropics, particularly in the Caribbean. With so much diversity in some environments, they have effectively partitioned the niches. In any given habitat there may be tree-trunk anoles, twig anoles and foliage anoles occupying various levels from low bushes to the canopy. Amazon Bark Anoles occur in primary and secondary forests and disturbed areas, where they are mainly associated with trunks and large branches.

References: 8, 62

PRICKLY KNOB-TAILED GECKO

Nephrurus asper

FAMILY:
Carphodactylidae
LOCATION:
Capella, Queensland, Australia
STATUS:
Least Concern, IUCN Red List

Geckos of the family Carphodactylidae – the only endemic Australian reptile family – are famous for their unusual and sometimes striking tails, including leaf- and carrot-shaped. For those that can break their tails, there is just one cleavage point at the base. A lost tail is forfeited completely.

Most species of knob-tailed geckos (genus *Nephrurus*) have fleshy tails that function as fat storages and can be discarded when the lizards are attacked. They all have a spherical knob on the tail-tip when original but regenerated tails lack knobs. Their function remains unclear but knob-tailed geckos have been observed to create a buzzing noise by vibrating their tail-tips in dry leaf litter.

The Prickly Knob-tailed Gecko and a small cohort of close relatives are unique among the world's geckos as the only ones unable to lose their tails. They lack specialised cleavage points in the vertebrae that allow a tail to break. Their insignificant tail with the peculiar ball at the tip is most charitably described as humorous and one author referred to it as 'pathetically abbreviated.' It would have little effect in wriggling to distract a predator, so when a tail is reduced to little more than a bump on a rump there is no point in its loss!

Prickly Knob-tailed Geckos live in rock outcrops, forests and woodlands in the dry tropics of Queensland. They are fully terrestrial, sheltering by day in burrows including those dug by other animals. At night they emerge to hunt insects, spiders, scorpions and smaller geckos.

If provoked, Prickly Knob-tailed Geckos raise their plump bodies high on slender limbs, arch their backs, rhythmically sway from side to side then leap at their aggressor with mouth agape uttering a harsh wheezing bark. To me that spectacular response is amusing, but to a small mammal it would be a startling act of defiance.

A colleague of mine found two geckos 1 metre (3 feet) apart while spotlighting. The look of wonder on her face was priceless as she knelt on hands and knees beholding the fat little creatures in her torch-beam as though they were a pair of aliens. They were the main subject of conversation for days afterwards.

With an oversized head, a body ornamented with rosettes of spines and a tail so small it borders on ridiculous, the Prickly Knob-tailed Gecko has followed an evolutionary path that is utterly divergent from our idea of what a typical lizard looks like. That is its unique appeal.

References: 9, 50

43

GULBARU LEAF-TAILED GECKO

Phyllurus gulbaru

FAMILY:
Carphodactylidae
LOCATION:
Hervey Range, Queensland, Australia
STATUS:
Endangered, IUCN Red List

On a steep rocky slope, fig roots grip granite boulders and thick vines reach for the canopy. It is a tiny patch of dry rainforest surrounded by grassy woodland and grazing country near the Queensland city of Townsville. A population of Gulbaru Leaf-tailed Gecko is isolated in this pocket, hiding in the crevices by day and foraging over the rock surfaces at night.

The geckos are restricted to fragmented stands of rainforest growing among boulders on rocky slopes and gullies. Each year their domain shrinks as late dry-season fires chip away at the edges of these stands. It is an ongoing problem that is threatening the species' existence.

This is one of several leaf-tailed gecko species that occupy tiny distributions in tropical and subtropical forests of eastern Australia. Their evolution into separate. narrowly endemic species, some separated from each other by less than 20 kilometres (12 miles), harks back to broad ancestral distributions across extensive forested habitats that have expanded, contracted and been fragmented by the ancient changing of climates. The evolution of species in isolation, some confined to single localities, clearly demonstrates the conservation value of seemingly insignificant stands of forest!

Gulbaru Leaf-tailed Geckos occur as six subpopulations with a combined coverage of only about 12 square kilometres (4.6 square miles). Based on genetic information, most of these subpopulations were probably already isolated from each other prior to European occupation, and had likely been separated for several thousand years. This speaks volumes about their inability to traverse barriers, even those that appear trivial, posed by unsuitable intervening drier habitat.

The granite boulders are flecked and dotted with minerals, and the geckos are marked over their backs and limbs to exactly match the background. They belong to a group of leaf-tailed geckos that have narrow, cylindrical tails rather than the broader leaf shapes of their relatives. Original tails are spiny and have white bands. Regenerated tails are smooth without bands. As luck would have it, we found one of each.

I never underestimate the extraordinary privilege of seeing first hand an Endangered species that is just hanging on in a fragmented and threatened habitat. The sweat down my neck and the mosquitos humming around my face were small inconveniences as I climbed the slope at night to visit one of these special pockets of rocky forest. As I departed and descended after photographing the geckos, I noticed how abruptly the forest gave way to the surrounding flammable grasses. I considered it a small blessing there was no smoke in the air tonight. As for tomorrow, or next month or next year, that is another matter.

Reference: 17

44

ROUGH-THROATED LEAF-TAILED GECKO

Saltuarius salebrosus

FAMILY:
Carphodactylidae
LOCATION:
Blackdown Tableland National Park,
Queensland, Australia
STATUS:
Least Concern, IUCN Red List

Among the great sandstone cliffs in the Central Queensland Highlands there are windblown caves. For countless generations the caves provided shelter for First Nations people. Under those rock overhangs they told stories, conducted rituals and they left their marks. Their paintings speak of countless generations of continuous occupation by the custodians of the land. The hand stencils outlined in ochre are particularly evocative for they relate to individual people.

Those stable, sheltered conditions, so suitable for painting and ceremonies under the rock ledges, are also ideal locations for geckos. At night Rough-throated Leaf-tailed Geckos emerge from narrow vertical crevices and caves to patrol the exposed sandstone faces. And as they have done for millennia they take up positions, with limbs splayed and clawed toes gripping securely, on and around those daubed and stencilled artworks.

The 18 species of leaf-tailed geckos occupy restricted pockets of habitat in eastern Australia between Sydney and Cape York Peninsula. Many have extremely small distributions in isolated rainforests and rock outcrops. The Rough-throated Leaf-tailed Gecko is an exception as it enjoys a relatively broad distribution. It is mainly a rock inhabitant with only limited occurrence in rainforests, occupying major outcrops, primarily of sandstone, in the southern interior of Queensland.

The sandstone boulders and rock faces of the Central Queensland Highlands are variegated with lichens. The leaf-tailed geckos match these patterns perfectly with the mottling and blotches on their backs. The disguise is enhanced by the broad splayed tail, with a pointed tip when original and blunt when regenerated, and the raised tubercles over the body, limbs and original tail. These combine to break its outline, disrupt its shadow and fuse it onto the rock. The geckos are common in the area and thanks to their large size they have a bright eyeshine in my head-torch.

It was a humbling and evocative experience to visit the indigenous art galleries at night and witness the combining of two impressive sights. Long before I visited the rock overhangs another person placed hand on rock and left an indelible mark. And now an extraordinary gecko, large and flamboyant, sits astride the stencilled adult hand. There are no lichens on those protected rock faces so the paintings look fresh, and the gecko stands out against the unblemished stone and the ochre artworks. It certainly demonstrates the size of the lizard, but it also speaks to me of a continuum, an overlaying of Australia's unique natural heritage and vibrant culture.

45

SRI LANKAN GOLDEN GECKO

Calodactylodes illingworthorum

FAMILY:
Gekkonidae
LOCATION:
Nilgala, Sri Lanka
STATUS:
Vulnerable, IUCN Red List

In a sheltered cave set among tropical woodlands, large geckos chatter loudly and scamper across the rock walls. By any reckoning, Sri Lankan Golden Geckos are impressive lizards. They have a head-and-body length of up to 10 centimetres (4 inches), long prehensile tails, broadly splayed fingers and toes and huge protruding eyes. Males also have a golden wash over their throats.

Endemic to south-eastern Sri Lanka's lowland dry zone, Sri Lankan Golden Geckos are extremely social, dwelling as colonies in granite caves and overhangs. A typical cave may harbour several discreet groups of three to eleven individuals, and a population of up to 50.

But it is the reproductive stakes where these geckos really stand out. Like other members of the family Gekkonidae, they lay a maximum of two eggs per clutch, which are soft when laid, with the shells becoming hard and brittle when they dry. And like some other family members, they 'glue' these eggs to firm substrates. Each round egg adhered to the wall has a flat contact surface, effectively forming a hemisphere. There, the eggs harden and stick.

Sri Lankan Golden Geckos glue their eggs communally to rock walls in caves, under archways and beneath ledges. Some egg clusters may comprise up to 100 eggs. Generations of geckos return to the same favoured sites, and over time many hundreds or thousands may be deposited. When each egg hatches, most of the eggshell falls, but a small disc of adhered shell remains. Over time, the prints of their eggs have accumulated to form massive 'egg scars'. Some scars cover nearly 1 square metre (11 square feet). It is quite possible that these sites, featuring a mix of old traces and newly laid eggs, could have been under continual use for several centuries and be still going strong.

This long-term site-fidelity may be triggered by the certainty of correct temperature and humidity, plus a proven inaccessibility to predators. And with prior hatching success rates, each gecko could be returning to lay at the precise site where it emerged from its own egg.

Communal egg-laying is well documented among many groups of lizards, with some geckos and skinks visiting sites over successive seasons (see page 162). But nothing quite matches the unique strategy that this gecko shares with its closest relative, the Indian Golden Gecko (*C. aureus*) of India's Eastern Ghats. Generations of lizards faithfully gluing ever-larger accumulations of eggs onto a rock face. That's really something different!

References: 23, 78

46

SOLOMON ISLANDS BENT-TOED GECKO

Cyrtodactylus salomonensis

FAMILY:

Gekkonidae

LOCATION:

Tenaru, Guadalcanal,
Solomon Islands

STATUS:

Near Threatened, IUCN Red List

There are more than 330 described species of bent-toed geckos spread in a giant arc from Solomon Islands in the western Pacific to the Himalayas. It is the third-largest genus among vertebrates, and taxonomists are adding more species to a growing list. It is a near-certainty that even more will be discovered.

Small wonder they are a confusing bunch to identify, and not too surprising that the stunning gecko I found clinging to a vine in rainforest on the island of Guadalcanal remained undescribed until ten years after I took its picture.

Bent-toed geckos are named for their angular digits, which are clawed, slender and somewhat bird-like with none of the adhesive pads present on many other geckos. This means they cannot climb smooth surfaces. However, they are sure-footed and agile on rough trunks, vines and rocks. Many species have prehensile tails and are capable of executing well-coordinated leaps between branches. They are particularly diverse where limestone and rainforest habitats combine. In fact virtually every isolated limestone outcrop across Asia appears to support endemic species.

They also reach some impressive sizes. The Solomon Islands Bent-toed Gecko attains a head-and-body length of about 14 centimetres (6 inches), and a total length of more than double that.

Within its rainforest habitat the Solomon Islands Bent-toed Gecko prefers the larger trees, with a particular preference for sheltering in the cavities that form in strangler figs. At night they can be found on trunks and buttresses or, like the gecko I was lucky to come across, walking with ease along suspended vines.

Solomon Islands Bent-toed Gecko lies at the extreme end of one of the most extensive reptile radiations in the world. Their evolution into a wealth of species on mountains, outcrops and islands stops in the Solomon Islands. None have penetrated further into the Pacific, where different lineages of geckos have been extremely successful colonising far flung islands. The Solomon Islands Bent-toed Gecko is at the end of the line.

References: 49, 56

47

TOKAY GECKO

Gekko gecko

FAMILY:

Gekkonidae

LOCATION:

Rinca Island, Indonesia

STATUS:

Least Concern, IUCN Red List

What's turquoise with orange spots, has jaws like a vice, eats geckos and loudly proclaims its own name? The answer of course is Tokay, the formidable giant gecko of forests, plantations and human dwellings across much of South-East Asia!

The sight of a monstrous colourful gecko up to 35 centimetres (14 inches) long prowling the walls is not easily forgotten. When I clambered onto a chair to catch one in Indonesia it attracted a quite a crowd who correctly guessed they were in for an amusing spectacle. As I grabbed the lizard, it turned its matchbox-sized head around, grasped my thumb and sunk its eyes down to increase the pressure. The crowd laughed, the gecko hung on and I laughed too, trying to save face. It crushed my thumb causing a fine bead of blood to form along each side of the nail. Two lessons I learned were: Don't grab Tokay Geckos in front of a crowd, and best not do it at all!

The peculiar and resonant call of the Tokay Gecko starts with a few bursts of mechanical clicks, then a pause followed by several explosive utterances of *toh-kay … toh-kay … toh-kay*. It is obvious that the lizard has acquired its name from the sound it makes. On occasions I have shared a cabin with one or two Tokay Geckos, and that loud, sharp call has dragged me out of a deep sleep.

As well as catching insects attracted to lights, Tokay Geckos are voracious predators of other 'normal-sized' geckos. They have also been recorded to eat snakes and small mammals. Because they are large, Tokay Geckos have sometimes been used to demonstrate how some geckos defy gravity on smooth vertical surfaces. The mechanics of each step can be observed more easily than on smaller geckos.

Contrary to popular belief they do not adhere by suction, but by molecular attraction called Van der Waals forces. Parallel ridges beneath each padded toe are covered with millions of microscopic hair-like structures called setae, which in turn have multiple branches at their tips. The forces are very weak at each setal tip, but with millions of points of contact on each foot the results are dramatic. As each toe is lifted, it is peeled up from the front to disengage the setae and Van der Waals forces.

I am an avid observer of the geckos clustered on walls under lights. They stalk moths, defend their turf and even dash across ceilings as though gravity does not exist. But I'm no longer tempted to jump on a chair when I see the giant turquoise ones with orange spots. I'm just happy to watch.

48

SABAH FLYING GECKO

Gekko rhacophorus

FAMILY:
Gekkonidae
LOCATION:
Mount Kinabalu, Sabah,
Borneo, Malaysia
STATUS:
Data Deficient, IUCN Red List

In the thick submontane forests on the flanks of Borneo's Mount Kinabalu, conditions are usually cool, wet and foggy, and the tree trunks are coated in a riot of mosses, lichens and orchids. Among those profuse epiphytes, the Sabah Flying Gecko blends into near invisibility thanks to its variegated mossy colours and elaborate outline of flaps and flanges. Occasionally these geckos also helpfully turn up on the walls of the national park buildings!

Despite its name, the Sabah Flying Gecko cannot fly and it is highly unlikely it can glide. It was once classified among a small group of true gliding geckos, which are equipped with flat, paddle-tipped tails and a loose parachute membrane along each flank enabling them to launch into the air and control their descent. Because its lateral flaps have a superficial resemblance, it was assumed the Sabah Flying Gecko could do the same thing.

I have been lucky to see quite a few of these extraordinary geckos in those mountain forests and I am certain their serrated outlines have evolved for camouflage rather than gliding. In keeping with its appearance, a more appropriate name would be 'Bornean Moss Gecko'.

When concealment fails and a gecko is confronted by potential danger, it puts Plan B into operation. By arching its back, raising its body off the substrate and widely gaping its mouth to reveal a bright red lining, it presents a formidable appearance that must cause many small predators to reflect whether tackling this apparition is worth the possible consequences.

This montane specialist has been recorded between 600–1,600 metres (2,000–5,250 feet) above sea level and is endemic to Borneo. Until recently it was believed to only occur around Mount Kinabalu in Sabah, but a recent record has been confirmed from Mount Penrissen in Sarawak.

When I walk in those wet mossy forests, structurally complex and crowded with so many strange plants jockeying for space, I can never help wondering what hidden eyes are watching me as I pass. It is a fair bet I have come close to some of these odd camouflaged geckos with their serrated outlines fusing them against the moss. But millennia of refined concealment never factored a head-torch and a reflected eyeshine. At night the orange glow of their eyes tells me exactly where they are hiding.

Reference: 65

49

FLAT-TAILED HOUSE GECKO

Hemidactylus platyurus

FAMILY:
Gekkonidae
LOCATION:
Siem Reap, Cambodia
STATUS:
Least Concern, IUCN Red List

The geckos start to gather around the lights at dusk. During the day they were grey with darker streaks but now they are ghostly pale and bleached of pattern. They have changed to their nocturnal colour. By dark, every outside light on the walls of my Cambodian lodging is patrolled by half a dozen or so Flat-tailed House Geckos.

These are true urban winners. The modified human environment offers plenty of opportunities to thrive. Gaps behind boards, wall hangings and fittings offer perfect daytime retreats while artificial lighting is a magnet for insect prey. Across South-East Asia, Flat-tailed House Geckos prefer buildings to their traditional trees and rocks.

As an avid lizard watcher, I am attracted to the spectacle of how they interact with each other and hunt their prey. Under those lights the geckos' lives are an open book. They have grown plump on the steady supply of moths and other insects drawn to the lights. I feel the tension as I watch one stalking an unsuspecting insect. Moving forward by increments, one slow, sure-footed step after the other and with eyes fixed on its prey, the gecko narrows the gap. It pauses, I freeze and watch, then the final fast lunge. I am not sure if I am barracking for the gecko or the hapless insect lured to the danger of artificial lighting, but I keenly feel the gecko's loss if it misses.

I notice one of the geckos is a bit different to the others. It has a forked tail. This is the result of an injury where the tail was damaged. The original tail is supported by a row of vertebrae featuring specialised, weakened breakage points. If grasped, the tail can be severed easily at one of these points. The broken tail wriggles, distracting a predator while the gecko escapes. When a new tail grows from the stump these vertebrae are replaced by a simple cartilage rod.

The growth of extra tails often results from an incomplete break on the original, triggering a new tail to sprout from the wound, which heals and the original remains. But often, as in this case, there are two regenerated tails. Perhaps they result from lacerations rather than a clean break on the site of the injury.

The fork-tailed gecko was a standout among its cohorts, and I kept an eye out for it each evening. There was one light it particularly favoured so it largely confined itself to part of one wall. I do not know if the other geckos may have also had their regular spots. To me they all looked the same.

50

SATANIC LEAF-TAILED GECKO

Uroplatus phantasticus

FAMILY:
Gekkonidae

LOCATION:
Analamazaotra-Perinet, Madagascar

STATUS:
Least Concern, IUCN Red List

Not all leaves that fall in a tropical rainforest land on the ground. Some end up snagged in tangles of vines and spider webs and accumulated in epiphytic plants. Suspended aggregations of leaf litter, decaying in the humid forest air, are effectively micro-ecosystems. For the Satanic Leaf-tailed Gecko, a slow-moving, sure-footed arboreal lizard that patrols the vines and branches at night, they are favoured daytime haunts.

There are more than 20 species of leaf-tailed geckos in Madagascar, belonging to the endemic genus *Uroplatus*. They are all masters of camouflage but they achieve their ends in different ways.

Some species have patterns resembling variegated mosses and lichens. Their frilled and flanged outlines fuse them shadowlessly against forest tree trunks so they are 'at one' with the textures and nearly invisible.

The Satanic Leaf-tailed Gecko, along with a cohort of close relatives, mimics dead, decaying leaves. With angular spines on the elbows and knees, raised peaks above the eyes and a splayed, asymmetrically notched tail it has all the hallmarks of a rotting, insect-ravaged leaf. It would take an extremely keen eye to spot it hunched among that dead suspended leaf litter.

When prolific taxonomist George Albert Boulenger named this species in 1888 it was based on a preserved specimen with no tail, lodged in the British Museum of Natural History. But even without that spectacular leaf-like appendage he was clearly intrigued by the unique appearance of this oddity. For he named it *phantasticus,* alluding to 'imaginary' or 'mythical.'

There were no field guides to the reptiles of Madagascar when I visited the island in the 1980s. Nor were there published images of much of the unique wildlife. Such was apparently the case with the Satanic Leaf-tailed Gecko. The only picture I had stored in my mind's eye, as I walked at night through alien forests with my head-torch, was the line drawing of the tail-less original specimen. Imagine my shock to suddenly encounter a live one, complete with a resplendent tail like a chewed dead leaf. Moments like these become life's treasured memories.

Dr Boulenger would never have seen a live specimen, nor visited its habitat in the low- to mid-altitude rainforests of eastern Madagascar. However, I am sure, had he spied a Satanic Leaf-tailed Gecko stalking slowly and cat like along a slender mossy vine, as I did that night, he would be more than satisfied that this odd creature is aptly named.

Reference: 14

51

BLUE-TAILED DAY GECKO

Phelsuma cepediana

FAMILY:
Gekkonidae
LOCATION:
Mahebourg, Mauritius
STATUS:
Least Concern, IUCN Red List

Bold, highly conspicuous colours come at a price. There needs to be a back-up plan because being noticed can be dangerous. Day geckos of the genus *Phelsuma* sport vivid hues akin to some tropical reef fish. Brilliant green and turquoise are common combinations, often emblazoned with scarlet spots or stripes. Those colours probably evolved so the lizards notice each other, but they can also attract potential predators. So their back-up is speed. They out-pace virtually everything that tries to catch them, myself included.

Typical geckos are nocturnal and it appears that the day geckos have secondarily evolved to be active during the day. In the process their eyes have become slightly smaller than those of other geckos, and their pupils are round rather than elliptic.

More than 50 species of these racy little beacons are shinning up and down tree trunks on islands across the western Indian Ocean, Andaman and Nicobar Islands and south-eastern Africa.

The Blue-tailed Day Gecko lives on the island of Mauritius where it thrives in parks, gardens and dry lowland forests that include a mix of native and exotic plants. It prefers palms and palm-like trees with smooth trunks and fronds, and sheltered crevices that hold water.

Like other geckos it feeds on insects, but it also takes nectar. It plays a critical role in the ecology of a rare endemic plant, *Roussea simplex*, the sole member of the family Rousseaceae. The Blue-tailed Day Gecko is possibly the only animal that acts as both pollinator and disperser. The gecko visits the flowers for nectar, and it laps up a gelatinous secretion produced by the fruit containing the minute seeds, then it disperses the seeds in its droppings.

Unfortunately an introduced species, the White-footed Ant (*Technomyrmex albipes*), also forages on the plant and has begun monopolising the nectar and fruit pulp. When it aggressively excludes the geckos, their visits virtually cease. Experiments have shown that if geckos are prevented from visiting, the plant's pollination and seed dispersal is impacted and the rare species is further threatened.

So while the Blue-tailed Day Gecko adds a gorgeous splash of colour to parks, garden trees and the dwindling areas of natural vegetation across much of Mauritius, there is an ongoing battle between the geckos and the ants to access and propagate one of the rarest plants on the island. This gecko is not just a pretty face!

References: 35, 36

52

NAMIB WEB-FOOTED GECKO

Pachydactylus rangei

FAMILY:
Gekkonidae
LOCATION:
Sesriem, Namibia
STATUS:
Least Concern, IUCN Red List

The Namib Desert features enormous wind-sculpted sand dunes, including some of the tallest in the world. Rain is a rarity, and most available moisture is delivered as fog. Some of the inhabitants of this challenging terrain have evolved extreme modifications to deal with the harsh conditions.

There is a gecko prowling the sands with fancy webbed feet. The Namib Web-footed Gecko mainly occupies the compacted windward faces of dunes, though it often forages at night on the nearby sandy flats.

Webbing between the toes of lizards is a rarity, and the Namib Web-footed Gecko is one of the few lizards in the world, and probably the only terrestrial species, to have them fully webbed front and rear.

It has been generally assumed those strange feet evolved to help traverse loose sand. Disperse the weight and reduce the hindrance of toes sinking in, like snowshoes. But it seems the webbed feet are mainly designed for digging.

Namib Web-footed Geckos shelter by day in burrows where the temperature is stable with little fluctuation year-round, despite extremes of heat and cold on the surface. The webbing is used during excavation, acting as shovels to remove and discard sand by the fistful. All limbs are used to dig, the front to loosen and excavate, and the rear to push sand backwards. There are plenty of lizard species around the world that dig burrows, but this is a unique set of equipment.

It has recently been revealed that Namib Web-footed Geckos are actually fluorescent, exhibiting some of the brightest fluorescence among terrestrial vertebrates. It is probably visible under moonlight, and appears as neon-green areas around the eyes and lower flanks. Those sites are strategically placed so they are largely concealed from predators above, such as birds and jackals, but highly conspicuous from the geckos' point of view.

Namib Web-footed Geckos live in an arid environment where virtually all moisture is acquired by licking condensed fog. It has been proposed, based on captive animals but not proven in the wild, that enhanced opportunities to meet up may enable them to lick moisture from each other's bodies.

I knew there was something very special about those Namib Web-footed Geckos when I found them in my torchlight. But I did not realise at the time just how extraordinary they were, these little lizards with feet like shovels, who locate each other in the moonlight by their neon-green fluorescence.

References: 64, 70

53

MOORISH GECKO

Tarentola mauritanica

FAMILY:
Phyllodactylidae
LOCATION:
Elba, Italy
STATUS:
Least Concern, IUCN Red List

From the train between Rome and Naples, I saw crops being harvested, villages and plenty of densely settled urban scenery, but not much of the evocative vistas for which Italy is famous. As the countryside rolled by I regarded it with only limited interest. Suddenly I had the flash of a lizard, perhaps a fat gecko. It was gone in a second. Was it real or a fancifully shaped mark on a fence? Nose to the window, I saw another, and I had certainty. Plump grey geckos were basking in the winter sunshine on fences and posts beside the railway line. Many of them in heavily built up areas.

When we arrived at Pompeii I lunged at one on the train station, missed it and made a spectacle of myself among the throng of alighting tourists. It was a week later during a walk on the island of Elba, where I saw Moorish Geckos closely on rock faces. Although mainly nocturnal they commonly bask during winter.

Moorish Geckos are particularly common in modified environments of southern Europe and northern Africa, and actually appear to prefer them to more natural areas. Their use of urban habitats has been studied in Rome where there are interesting dynamics between Moorish Geckos and Turkish Geckos (*Hemidactylus turcicus*). Both are common but each greatly outnumbers the other at different sites. Moorish Geckos prevail on damaged and ancient walls while Turkish Geckos have the upper hand on more modern buildings. Moorish Geckos are believed to have a longer history of occupation in the area.

The generic name *Tarentola* is derived from the Italian town Taranta Peligna. The word 'tarantula' is named for the same place, though originally applied to a wolf spider. Residents wrongly believed both the geckos and spiders were dangerously venomous. Being covered with large, very prominent tubercles, having strange feet and moving mainly at night probably did little to lift the geckos' reputation.

It is hardly surprising that Moorish Geckos have adapted well to human habitation. They overlap with more than two thousand years of intensive agriculture, erection of stone walls and the building of cities, and of course the construction of roads and movement of goods provide dispersal opportunities. Only versatile species can persist and actually thrive on grimy fences beside railway lines where they occasionally catch the surprised eye of a passing herpetologist.

Reference: 47

54

CRESTED GECKO

Correlophus ciliatus

FAMILY:

Diplodactylidae

LOCATION:

Parc Provincial de la Rivière Bleue,
New Caledonia

STATUS:

Vulnerable, IUCN Red List

New Caledonia's Crested Gecko, one of the world's most spectacular geckos, has a chequered history. Feared extinct as recently as 1993, it is now among the most popular pets in the world of reptile keeping.

When it was described in 1866 the type locality, listed as simply 'Nouvelle-Caledonie', shed no light on its provenance. It was considered common before it vanished for nearly a century. There was a concern the lizard may have been wiped out, but in 1994 it was rediscovered on Ile des Pins (Isle of Pines), about 50 kilometres (30 miles) off the southern tip of New Caledonia.

The Crested Gecko occupies warm, moist forests in widely scattered localities across the southern mainland of New Caledonia (Grande Terre), and it has recently been discovered in the north too. Populations possibly persist in intervening forested areas.

It is a sight to behold, with an impressive crest of spines above the eyes that flare out along each side of the head and onto the neck, webbed hindlimbs, broad digits and a flattened paddle-shaped tail-tip equipped with adhesive lamellae bristling with microscopic hair-like structures called setae. Like nearly all other geckos it can readily lose its tail, but it rarely grows a replacement. Once the whole tail is lost, it is gone for good.

Crested Geckos are agile and well designed for life in the trees. They forage mainly on slender branches of the canopy rather than the main tree trunks, and sleep by day among twigs and presumably arboreal debris.

Early field observations suggested Crested Geckos were only seen after rain. That accords with the animal pictured, which was one of three specimens that I found foraging on saplings and tree limbs in dense forest during and after torrential downpours.

Crested Geckos are now extremely popular pets. A quick browse of the internet reveals numerous care sheets and other husbandry guides, and there are plenty of Crested Geckos advertised for sale.

Despite its popularity, it is a threatened species in the wild. Introduced deer and pigs are degrading habitat and populations at lower elevations must also deal with the introduced Little Fire Ant (*Wasmannia auropunctata*). And those lizards inhabiting accessible locations may find themselves caught up in the illegal collecting and trafficking trade.

It has been an odd journey for the Crested Gecko, from early discovery to presumed extinction, to popular pet, to real threats in the wild state.

References: 9, 11, 74

55

LARGE-SCALED CHAMELEON GECKO

Eurydactylodes symmetricus

FAMILY:

Diplodactylidae

LOCATION:

Parc Provincial de la Rivière Bleue,
New Caledonia

STATUS:

Endangered, IUCN Red List

The Large-scaled Chameleon Gecko doesn't fit the usual descriptions. The simple line 'head and body covered with small granular scales' refers broadly to virtually all of more than 1,900 described gecko species in the world. There are plenty of variations including tubercles, spines and rosettes, but those words or similar appear in countless books and journals. This gecko is an exception. It is the most extreme of four endemic New Caledonian chameleon geckos in the genus *Eurydactylodes,* known for their enlarged plate-like head shields. These are not your standard geckos.

Chameleon geckos live in forested areas over most of New Caledonia, with the Large-scaled Chameleon Gecko occurring in the south. They are all arboreal, living mainly among leaves and twigs rather than trunks and larger stems. They even use these as daytime retreats and perches, clinging exposed rather than seeking enclosed shelters. They are also partly diurnal.

These odd little lizards feature a peculiar bright yellow groove of unscaled skin between the jaw and the ear. It is an obvious feature and its function has not yet been confirmed, but the mouth-lining is the same conspicuous colour and it has been proposed that this forms part of a gaping display.

When threatened by predators, chameleon geckos have an interesting strategy at their disposal. They can smear a smelly sticky fluid, or even squirt it up to 50 centimetres (20 inches), from ducts within the tail. This important mechanism may explain why broken and regrown tails are uncommon. Curiously, the only other geckos in the world with this ability are the striped, jewelled and spiny-tailed geckos in the genus *Strophurus,* all endemic to Australia (see page 128). Many of these also have bright coloured mouths, and frequently select exposed stems and foliage as diurnal resting sites, often in bright sunshine.

Within the family Diplodactylidae, chameleon geckos are more closely related to other species that have radiated throughout New Caledonia than they are to their Australian counterparts. So their coloured mouth-linings, diurnal perching sites and the unique delivery of a repellent substance have almost certainly evolved separately in the two groups.

The Large-scaled Chameleon Gecko is considered Endangered, facing threats from expanding nickel mines, clearing of native habitat for exotic plantations and impacts from the introduced Little Fire Ant (*Wasmannia auropunctata*), The unusual lizard is also desirable among reptile keepers, exposing it to illegal trafficking. Fortunately it inhabits some protected sites including Parc Provincial de la Rivière Bleue.

References: 10, 11

56

GARGOYLE GECKO

Rhacodactylus auriculatus

FAMILY:
Diplodactylidae
LOCATION:
Parc Provincial de la Rivière Bleue,
New Caledonia
STATUS:
Least Concern, IUCN Red List

There is a rugged land with unique vegetation where the complex ecosystems feature lizards as apex predators. A land where the only native mammals are bats, but where an array of giant geckos and skinks prey on nestling birds, large insects and each other. That land is New Caledonia.

Geckos of New Caledonia belong to two different families. The cosmopolitan Gekkonidae is well represented across far-flung parts of the Pacific, and the regionally endemic Diplodactylidae includes many species in Australia and New Zealand. It is this family that contains New Caledonia's most unusual geckos.

The Gargoyle Gecko, also sometimes called the Knob-headed Gecko and New Caledonian Bumpy Gecko, is named for the gnarled topography of its large head. And with a head-and-body length of 12 centimetres (5 inches), a plump body and slender tail it is quite a sight to behold.

It is endemic to the southern third of New Caledonia, particularly where the habitat is dominated by vegetation called maquis. New Caledonian maquis is defined by its low heath-like plants with tough, thickened leaves and stunted *Araucaria* trees growing on infertile soils rich in minerals, particularly nickel. The gecko is also found in humid forest, mainly along the edges where it abuts maquis.

New Caledonia's lizard communities are relatively simple, comprising only skinks and geckos. But like typical reptile communities, they achieve a status quo through niche partitioning to reduce competition for food and resources. Different diets, shelter sites and activity periods reduce overlap. Gargoyle Gecko is one of the largest lizards in its habitat. This alone separates it from most other local species. It is usually found in shrubs and low trees, effectively isolating it from other large geckos in the area that prefer higher sites in taller vegetation.

The lizard is believed to have one of the broadest diets among geckos globally. As well as the usual range of arthropods, it eats smaller geckos and skinks, snails, flowers, leaves and sap. It has been suggested that its unusually long, fang-like teeth are a modification to deal with lizard prey.

Because of its unusual appearance including numerous attractive colour variants, the Gargoyle Gecko is popular with reptile husbandry enthusiasts and breeds well in captivity. However, it has a powerful bite when defending itself, and would be quite formidable when dealing with prey. With those long blade-like teeth, this is not a gecko to be messed with.

References: 11, 77

57

NEW ZEALAND JEWELLED GECKO

Naultinus gemmeus

FAMILY:
Diplodactylidae

LOCATION:
Pōhatu/Flea Bay, Banks Peninsula,
New Zealand

STATUS:
Endangered, IUCN Red List

A brisk cool wind off the southern ocean cut through my shirt but at least the sun was out and the temperature was about 14°C (57°F). That thin watery sunshine created perfect basking conditions for cool-adapted New Zealand Jewelled Geckos.

They were living in bushes called *Coprosma*. Each well-spaced shrub consisted of a compact lattice of tough, interlocked stems and a dense matrix of small round leaves with foliage so thick it was like a green wall that could barely be penetrated.

The bright green geckos inhabiting these bushes live life in the slow lane. They emerge to bask partly concealed among the outer foliage, eat berries and snap flies, and with a few slow casual steps can retreat into the dense matrix and be unreachable. For me it was a thrill just to see these iconic lizards basking.

There are about 20 named species of New Zealand geckos, and as many more under investigation and pending descriptions. They are a unique bunch. Excepting two New Caledonian species they are the only live-bearing geckos in the world. A feature of all geckos is their small clutch size, almost invariably fixed at two, occasionally one. New Zealand geckos follow this trend, giving birth to fully-formed twins. They are generally long-lived, sometimes exceeding 50 years of age!

New Zealand's live-bearing geckos have responded to extended isolation in a cold climate. Instead of laying eggs they retain embryos in their bodies. With a gestation period of 3–14 months, they regulate the temperature of developing embryos by basking or moving into contact with sun-warmed surfaces. Along with nearly all of New Zealand's skinks they form the largest assemblage of live-bearing lizards in the world.

The eight species of green geckos are also unusual in being diurnal. They are all arboreal, but rather than living on trunks and large branches they inhabit stems and foliage. Favoured plants include manuka, kanuka and *Coprosma*.

Although they are protected by law, New Zealand's green geckos are highly desired by collectors, and poaching for the wildlife trade is a serious concern. They can fetch high prices in the thousands of dollars. And then there is a menagerie of exotic mammals to deal with. These include rats, mice, stoats, weasels, ferrets, possums and cats. And with the loss of more than two-thirds of forest cover, it is not surprising that green geckos are of conservation concern.

In the face of all these problems I've got my fingers crossed that the New Zealand Jewelled Geckos I watched on the Banks Peninsula are still there today, soaking up that weak, watery sunshine.

Reference: 88

58

PRETTY GECKO

Diplodactylus pulcher

FAMILY:

Diplodactylidae

LOCATION:

Johnson Rocks, Western Australia

STATUS:

Least Concern, IUCN Red List

Pretty Geckos are well named thanks to their attractive and conspicuous dorsal patterns. The lizard pictured with the irregular blotches is probably more typical for the species but the plain straight dorsal stripe is not uncommon. Curiously it is just one of several terrestrial gecko species exhibiting starkly different blotched and striped colour forms.

Pretty Geckos have tapped into an abundant food resource. They are termite specialists. Countless billions of these insects are critical elements as fodder and as structural architects in semiarid and arid ecosystems across Australia. All of those mandibles munching so many tonnes of wood and other tough plant material, processing it by harnessing protozoa and enzymes to break down cellulose, and ultimately converting it into the easily digestible little parcels of protein we all know and love.

Specialising on termites has obvious benefits so it is not surprising that many insectivores rely on them as their primary or sole food. Termites are abundant and they occur as a clumped or concentrated source. Small nocturnal lizards can find termites relatively easily and they can usually eat their fill. Several other terrestrial geckos in arid zones of Australia and skinks in Africa share similar narrow diets comprising termites.

Pretty Geckos occur widely over the southern interior of Western Australia, particularly where mulga and other hardy plants grow on heavy reddish soils. It never ceases to amaze me, when travelling in these areas during hot weather, how small and apparently delicate lizards are concealed somewhere under a shimmering heat haze.

Like several other arid-zone terrestrial geckos, including a number of other termite specialists, Pretty Geckos often shelter by day in the vertical shafts of spider holes, where they are thermally buffered from extreme temperatures at ground level. However in mild weather Pretty Geckos often hide under surface objects such as small stones. So when turning a slab adjacent to a remote granite outcrop I would not consider it unusual to find a Pretty Gecko sheltering beneath. The big surprise was to find three, including these striped and blotched forms sheltering together.

Reference: 61

59

NORTHERN MARBLED VELVET GECKO

Oedura marmorata

FAMILY:

Diplodactylidae

LOCATION:

Adelaide River area,
Northern Territory, Australia

STATUS:

Least Concern, IUCN Red List

Velvet geckos are named for their smooth velvety skin-surface and even, uniform scales. There are no bumps, spines or tubercles on their backs and limbs. Adults generally have complex patterns of bands, spots or blotches, or a combination of these. Juveniles are usually more simply patterned with highly contrasting spots or blocks of colour.

The 19 named species of *Oedura* live in woodlands and rocky habitats across most of Australia with the highest diversity in the seasonally dry tropical north. Some are fussy, living exclusively among rocks, while others prefer dead timber with loose bark. Several species simply don't care, provided there is a thermally secure narrow gap in rock or wood.

The Northern Marbled Velvet Gecko is a rock inhabitant, representing the genus on gorges and outcrops over much of the Northern Territory's 'Top End'. Prime habitats include the towering sandstone pillars of Litchfield National Park and the sheer rock walls and boulders of Nitmiluk (Katherine Gorge) National Park.

The lizards are abundant in suitable areas, where at night they prowl sure-footed across the rock faces and on the immediate adjacent ground. Sometimes it is their disturbance of the crisp dry leaf litter that gives them away.

There is a common trend among obligate rock-inhabiting lizards that span wide distributions. When the rock habitats are discontinuous, the various populations tend to look a bit different. That appears to be the case with the Northern Marbled Velvet Gecko, where broad expanses of tropical savannah woodlands act as barriers to dispersal. Subtle differences are apparent among populations.

Those from Litchfield and nearby Adelaide River are renowned for the intensity of their colour, particularly their vibrant lemon-yellow spots. Given their superb appearance it seems appropriate that they dwell among such scenic grandeur.

I am not one to judge whether any lizards are 'better' than others. To me all small brown lizards are marvels in their own right. However, having clambered up rock ridges at night and seen them in my spotlight, I cannot deny how those geckos with the bright yellow pattern made a deep and lasting impression on me.

60

GOLDEN-EYED GECKO

Strophurus trux

FAMILY:

Diplodactylidae

LOCATION:

Marlborough area, Queensland,
Australia

STATUS:

Not evaluated but recommended as
Vulnerable, IUCN Red List

Not much is known about the Golden-eyed Gecko, but it is assumed to be similar in general habits to its close relatives, the spiny-tailed, jewelled, and phasmid geckos.

Like other species of *Strophurus* it is probably a tail-squirter. When harassed these geckos can eject sticky repellent treacle-like fluid from ducts within the upper surface of the tail. This can be dribbled, smeared or forcibly ejected as a fine jet up to 1 metre (3 feet). On contact with air it quickly dries to cobweb-like filaments. It has an unpleasant odour resembling crushed legume seeds and can cause severe eye irritation. The only other tail-squirting lizards are the chameleon geckos of New Caledonia. They appear to have independently evolved this unique and highly unusual defence mechanism (see page 118).

The 20 described species in the genus *Strophurus* fall broadly into two behavioural groups. The arboreal members inhabit shrubs and other low woody vegetation featuring foliage cover, twigs and slender branches. They shelter in hollows, behind bark or cling to exposed branches, sometimes in full sunlight. The tussock inhabitants are grass specialists occupying clumps of spinifex, and occasionally sedges. Both groups use their padded toes to grip slender perches such as sticks and stems.

The Golden-eyed Gecko is closely associated with spinifex (*Triodia mitchelli*), a hummock grass comprising a dense lattice of slender spiny foliage. It has only ever been found by spotlighting, either on spinifex or on closely associated dead shrubs and other vegetation. It is restricted to a tiny area of steep rocky slopes near the Queensland town of Marlborough. Spinifex is highly flammable and significant portions of occupied habitat have been burned over the past 15 years.

In its original 2017 description it was recommended that the Golden-eyed Gecko be listed as Vulnerable under the IUCN Red List criteria. It is known from such a restricted area, essentially a single locality, and is likely to experience extreme fluctuations in area of occupancy and number of mature animals.

With landowner permission I trekked up to one of the few accessible sites in stiflingly hot December weather with naturalist friends Angus Emmott and Ross Coupland. It took several climbs, two nights of spotlighting and near heatstroke conditions before Ross finally found one on a dead woody stem protruding from spinifex. As it moved back into the vegetation I pondered how much there is to learn about this little lizard, tucked away on its remote, fire-prone stony slopes.

References: 17, 87, 95

61

BRIDLED FOREST GECKO

Gonatodes humeralis

FAMILY:

Sphaerodactylidae

LOCATION:

Rio Napo, Amazon Basin, Ecuador

STATUS:

Least Concern, IUCN Red List

It is often a relief to take time out from fieldwork in tropical rainforest. Remove the muddy boots, rest the feet and have a cup of tea. And while seated near the doorway of my bungalow nestled in pristine forest of Ecuador's Amazon Basin, I could watch a couple of geckos in relative comfort.

Unlike most geckos, Neotropical geckos of the genus *Gonatodes* are diurnal. They are swift, alert and, in keeping with many other day geckos including *Phelsuma* in the eastern Indian Ocean (see page 110), have evolved round rather than vertically elliptic pupils. There are more than 30 named species in tropical Central and South America and islands of the Caribbean. They sometimes bask in direct sunlight, but generally prefer semi-shaded areas. Males are territorial, typically living on tree trunks in the presence of one or two females.

Determining the sex of most geckos is usually easy. The males' testes form bulges on either side of the tail-base, though it sometimes requires a close look to see them. For most *Gonatodes*, the process is usually even simpler. The brightly coloured males look completely different from the drably marked females.

Male *Gonatodes* are advertising billboards, broadcasting themselves as holders of territories they are prepared to defend against intruders, and as viable mates for local females. Male Bridled Forest Geckos are normally fairly drab, but the sight of a rival nearby can trigger a dramatic and spontaneous display of brilliant red, yellow and sometimes purple. This may be accompanied by tail waving and body arching. If these fail, attacks or chases around the trunks and buttresses of trees may ensue.

Females are coloured a cryptic grey with darker blotches, and they appear incapable of colour change. As the carriers of eggs, it makes sense that they draw less attention to themselves than the conspicuous, risk-taking males with territories to defend and mates to woo.

Just one egg is laid per clutch but large numbers of eggs are sometimes found together, suggesting repeated visits to a site or its communal use by numerous females.

The range of the Bridled Forest Gecko extends across the vast Amazon and Orinoco Basins. It is one of those fortunate species that tolerates some disturbance, extending its habitat use beyond tree trunks in old-growth forests to walls, fenceposts, roof thatching and other human structures. That includes the support posts of my bungalow. So it turned out that the little lizards I was watching were not different species as I first thought, but a pair of Bridled Forest Geckos.

References: 8, 62

62

WESTERN BANDED GECKO

Coleonyx variegatus

FAMILY:
Eublepharidae
LOCATION:
Tortolita Mountains, Arizona, United States
STATUS:
Least Concern, IUCN Red List

The towering silhouettes of giant saguaro cacti loom tall, fading against the darkening sky as the last orange blush of sunset grows faint on the horizon. After a hot summer day of around 40°C (104°F) in the Sonoran Desert, the sand is cooling and things are starting to move.

A small gecko, freshly emerged from its insulated cavity beneath the ground, takes careful steps among the detritus of a fallen cactus. As I look closely, focusing to take a picture, it does something that I have never seen a gecko do before. It blinks.

The Western Banded Gecko belongs to a family unique among the world's geckos. The 44 species of eyelid geckos are the only ones that can close their eyes. They have moveable eyelids. All other geckos, around 1,850 described species, have fixed, immoveable transparent spectacles capping their eyes. Eyelid geckos also share slender toes lacking any adhesive pads, and nearly all are terrestrial.

The family has an intriguing, patchy global distribution in Central to North America, Africa, the Middle East, Japan and South-East Asia. It suggests the modern distribution of eyelid geckos is a relic, a fragmented and smaller version of what it once was.

The Western Banded Gecko is one of eight species occurring in the Americas. It lives in deserts from the south-western United States to western Mexico, often in harsh, sparsely vegetated environments. At least five different colour patterns exist within this broad range, each named and sometimes treated as a subspecies. The lizard pictured is sometimes called the Tucson Banded Gecko.

It is curious that this inoffensive, fragile-looking, soft-skinned little lizard, with a head-and-body length of just over 7 centimetres (3 inches), should be maligned and generate much fear. Yet historically that has been the case.

The Seri People of Mexico believed it to have highly toxic flesh and blood, causing fatal lung illness if touched, and that contact also results in one's flesh falling off. Even worse, if a person lies down where a gecko is buried, its silhouette will burn into the victim's body.

Such fears may stem from the lizards' habit of walking with the tail raised, possibly mimicking a scorpion. Or they may result from its defensive response of twisting and sinuously undulating the tail. We may never know how these beliefs originated, but they are the stuff of nightmares!

After I took my pictures, I picked the little gecko up and had a closer look before replacing it to resume its nocturnal foraging. I am still alive and fortunately I have all of my skin.

References: 9, 32, 81

63

LINED WORM-LIZARD

Aprasia striolata

FAMILY:

Pygopodidae

LOCATION:

Casterton area, Victoria, Australia

STATUS:

Least Concern, IUCN Red List

On a mild spring morning I was on a slope in far western Victoria with fellow herpetologist Mike Swan, carefully turning semi-embedded stones that I had wanted to look under for more than two years. There were some ants present, with their tunnels exposed beneath the stones. And there were Lined Worm-Lizards, the very animals I had travelled to see.

Timing is important, as worm-lizards mainly use these shelter sites during cool weather. At hot times of year they seek more insulated places, probably deeper down and under thick debris. So after delays from work and field commitments, Covid-19 lockdowns and travel restrictions we were at the right place and the right time in perfect weather.

More than a dozen species of worm-lizards occur mainly across southern Australia, on sandy soils in semi-arid and well-drained habitats. They live under partially embedded stones, stumps and tree roots, often in association with the galleries of ants.

It is difficult to believe at first glance, that these smooth and shiny, blunt-snouted and short-tailed burrowing lizards are closely related to geckos. They look nothing like them, yet the relationship between the family Pygopodidae and geckos is universally acknowledged.

They share key features, including a broad flat tongue used to wipe the clear spectacles covering their eyes. They have a voice in the form of a high-pitched squeak, sometimes uttered when under stress and possibly also during social interactions, and they lay a fixed clutch size of two eggs. Though appearing limbless, all species retain vestigial hindlimbs in the form of scaly flaps or spurs. Some herpetologists even regard pygopodid lizards as essentially near-limbless, snake-like geckos.

Over 95 per cent of the worm-lizards' diets comprise larvae and pupae of ants. A feeding observation of one species that may apply to others suggests infrequent 'binge-feeding' within the food-rich ant nests. A large number of ant larvae were gobbled up in mere seconds. The small soft prey are swallowed whole without chewing and as a result worm-lizards have greatly reduced dentition.

It is critically important that semi-embedded stones be replaced in their precise original position if future generations of ants and lizards are to continue using them. Turning them is like opening presents, except there is never a guaranteed prize. So when I lifted a stone and encountered a limbless lizard with sharp black stripes curled underneath, I knew this was worth more than any wrapped gift.

References: 39, 91

64

BURTON'S SNAKE-LIZARD

Lialis burtonis

FAMILY:

Pygopodidae

LOCATIONS:

Top left: Mount Isa, Queensland, Australia
Top right: Tieri, Queensland, Australia
Bottom left: Nappa Merri Station, Queensland, Australia
Bottom right: Brisbane, Queensland, Australia

STATUS:

Least Concern, IUCN Red List

I will never forget peering into a clump of spinifex in Victoria's semi-arid mallee region and seeing the striped body and acutely wedge-shaped snout of a Burton's Snake-Lizard. This early teenage experience was my first live encounter.

Burton's Snake-Lizard is the most widespread Australian terrestrial reptile. It occurs over most of the continent, and through the Torres Strait Islands to southern New Guinea. With the exception of some southern areas, wetlands and closed moist forests, Burton's Snake-Lizards can turn up in bushland just about anywhere.

It also has one of the most specialised diets of Australian lizards. It is an exclusive lizard predator. The Burton's Snake-Lizard has tapped into a diverse and abundant resource, and in doing so it has evolved convergently with Australia's small venomous snakes. Most of these also exploit lizards as a sole food source.

Like other pygopodid lizards, the Burton's Snake-Lizard is closely related to geckos, but ecologically, it is a de facto snake! This ambush hunter feeds mainly on surface-active lizards, lying motionless ready to snatch prey with a sideways swipe of the pointed snout.

As a non-venomous predator, the Burton's Snake-Lizard has evolved effective ways to dispatch and swallow food. It has a uniquely hinged skull with a flexible joint about level with the eyes, allowing the tips of its jaws to meet, fully encircling and suffocating the victim. And its fine, pointed teeth are backward-curved and hinged, folding back when pushed from the front but locking in an erect position when pressured from behind. With a secure grip, it can manipulate prey and swallow it head first. Just like a small snake.

Burton's Snake-Lizards are also among Australia's most variable lizards. The bewildering array of colour forms include cream and patternless, through grey to bright yellow and brick red. Some have lines of dots or boldly contrasting stripes. While several quite different forms are usually present in any given location, there are some common local trends. Bold narrow stripes along the body are common in spinifex deserts, and broad black-and-white stripes on the face and forebody prevail in northern tropical woodlands.

I have seen Burton's Snake-Lizards crossing roads at night, vanishing into thick grasses and lying motionless waiting to ambush lizards. But the most memorable encounter was that first one. As I embarked on one of my early passionate quests to find lizards, it was there looking back at me through the spiny latticed stems of that spinifex clump.

References: 57, 95

65

ARMADILLO LIZARD

Ouroborus cataphractus

FAMILY:
Cordylidae
LOCATION:
Lambert's Bay, South Africa
STATUS:
Least Concern, IUCN Red List

As a tail-biting ring of spines, the Armadillo Lizard is one of a kind. When researchers introduced me to a colony of these oddballs in the Western Cape of South Africa, I found myself playing an unusual waiting game with an Armadillo Lizard and an Angulate Tortoise (*Chersina angulata*)!

These communal lizards share horizontal rock crevices, usually sandstone with an accompanying ledge suitable for basking. A colony may number more than 30, typically including several adult males in larger groups, and only one adult male in communities of nine or fewer. Although broadly insectivorous, they rely strongly on the Southern Harvester Termites (*Microhodotermes viator*) throughout the year.

Armoured with rings of tough scales, each projecting out as a large spine, they wedge themselves tightly in crevices and there is little chance anything can dislodge them. But they also have another quite extraordinary defence. If captured they bite their own tail and hold on tightly, effectively forming an impenetrable spiny ring or ball. A difficult shape to swallow! This response is reported from just about every encounter. The animal I handled did so immediately.

Armadillo Lizards are named after those armoured mammals, the armadillos, which also roll into balls as a defence. And the generic name derives from the mythological Greek Ouroborus, a dragon or snake illustrated eating its own tail.

I was thrilled to witness this phenomenon, and took photos as the lizard firmly gripped its tail. I also wanted portraits of the animal in a more conventional pose but it could not be induced to release its grasp so I decided to wait. It would let go eventually.

Angulate Tortoises are extremely common in the area and I saw one walking close by. I decided to photograph it while I waited but the tortoise promptly retreated into its shell. This placed me in the odd situation of crouching before a lizard biting its tail and a tortoise with its head pulled in. My patience eventually paid off on both counts.

Armadillo Lizards are widespread in the pet trade, and wild populations have been extensively exploited. Because they live communally, overcollecting can potentially wipe out whole colonies. Low fecundity and slow dispersal rates also inhibit their re-establishment at targeted sites. They were once listed as Vulnerable on the IUCN Red List, but because of their relative abundance across western South Africa they are now considered as Least Concern. The story of the curious Armadillo Lizard remains a gripping tale.

References: 53, 76

66

AUGRABIES FLAT LIZARD

Platysaurus broadleyi

FAMILY:
Cordylidae

LOCATION:
Augrabies Falls National Park,
South Africa

STATUS:
Least Concern, IUCN Red List

Dozens of flat lizards lie basking like vivid beacons on the rocks. The males are gaudily coloured and multi-hued, as though daubed by an overzealous child with a bright new set of paints. The Augrabies Flat Lizards inhabiting the smooth granites of South Africa's Augrabies Falls National Park have a narrow distribution but they can accurately be described as 'extremely abundant'. The park receives plenty of visitors and the paths are well trodden, so the lizards pay little heed to human traffic.

Flat lizards are aptly named. Their heads and bodies are pancake-flat, allowing easy access into narrow crevices in rocks and gaps under slabs. There are about 16 described species occupying rocky habitats, typically granite, gneiss and sandstone, across southern Africa. Species rarely overlap, with each separated by intervening terrain of unsuitable habitat. With their gaudy hues and distinctive shape, flat lizards have been affectionately described as 'depressed dandies'.

Large numbers of Augrabies Flat Lizards may pack themselves into a suitable crevice. More than 100 have been recorded in a single shelter site. These big aggregations tend to be strongly male-biased. By day the rocks are festooned with them. The boldly coloured male and brown female Augrabies Flat Lizards lie with long limbs splayed wide on the vertical and horizontal surfaces edging the walking trails.

Their colours appear lurid to us, but they also include visual cues perceived by the lizards that are invisible to humans. Male flat lizards view each other in the ultraviolet spectrum, which we cannot see, to assess each other's fitness. That may determine whether or not to mount a challenge and start a fight.

Some of the brightest colours are actually on the belly, concealed until strategically exposed. Rivals circle each other and reveal those ventral hues by tilting sideways and expanding their throats to present the ultraviolet signals. Though critical as a means of communication and determining status, being conspicuous is expensive and coloured males may be at greater risk of predation than the more concealed females. Among the most significant predators are Rock Kestrels.

Augrabies Flat Lizards aggregate around two main food sources. They feed on fallen fruits of the Namaqua Fig, and in summer they gorge on ephemeral plumes of blackflies, gathering to execute spectacular leaps and mid-air pirouettes to snatch the insects on the wing.

I have seen a lot of reptiles in many places, but walking among the kaleidoscope of abundant lizards at Augrabies Falls was a novel and unforgettable experience for me.

References: 15, 73, 80

67

FOUR-LINED GIRDLED LIZARD

Zonosaurus quadrilineatus

FAMILY:
Gerrhosauridae

LOCATION:
Ifaty, Madagascar

STATUS:
Vulnerable, IUCN Red List

On the edges of a narrow sandy track in the spiny forests of Madagascar's arid south-west, large shiny lizards dart for the cover of shrubs and fallen logs when approached. They are common and conspicuous as they dash across open ground and move among the leaf litter. These sleek alert lizards, with a head-and-body length of 16 centimetres (6 inches), are called Four-lined Girdled Lizards. They belong to a family of nearly 40 species found throughout sub-Saharan Africa and Madagascar.

Girdled lizards are readily identified from superficially similar and globally widespread skinks by their distinctive scales, which are arranged transversely side by side instead of in oblique rows, and by the prominent fold with small granular scales along each side of the body.

Girdled lizards are widespread in Madagascar, with about 20 species found over the island. Some have highly restricted ranges and face a variety of threats while other highly adaptable species occur across a range of habitats including disturbed urban areas. Some even forage for scraps in and around houses and largely ignore the human occupants.

There is a trend among diurnal, sun-loving lizards that eat arthropods. The families with alert, upright postures such as iguanas and dragons are often 'sit and wait' predators while lizards with a lower, more head-down posture, including skinks and lacertids are more likely to be active foragers. There are many exceptions on both sides, but the trend is fairly clear. Girdled lizards conform to this second group, and as they move through the landscape they investigate burrows, check the leaf litter for scent or movement, and examine cavities in and under logs in their hunt for invertebrates.

The Four-lined Girdled Lizard has an extremely limited range in south-western Madagascar, in arid habitats with red sandy soils and an array of unique spiny vegetation including many succulents. The rate of habitat destruction in the area is severe and the species' distribution is extremely fragmented. It is thought to now occur in as few as ten locations. Small wonder it is of grave conservation concern and considered Vulnerable.

68

PREHENSILE-TAILED SKINK

Corucia zebrata

FAMILY:
Scincidae

LOCATION:
Ngattokae Island,
Solomon Islands

STATUS:
Near Threatened, IUCN Red List

When two colleagues and I looked up at a giant fig tree with our torch-beams in the western province of the Solomon Islands we saw a Prehensile-tailed Skink. We knew it was a lucky find, but did not quite realise the extent of that luck! Subsequent radiotelemetry work has revealed that these exclusively arboreal lizards spend most of their time in the canopy and conventional surveys for them, undertaken by searching tree trunks from ground level, have a low chance of success.

Prehensile-tailed Skinks live in the larger forest trees and move between connected canopies. That includes giant figs, those wonderfully complex structures with labyrinthine cavities and tunnels, aerial roots and giant buttresses with their attendant climbing and epiphytic plants.

Prehensile-tailed Skinks live in forests across the Solomon Islands and Bougainville. Growing to a total length of more than 80 centimetres (30 inches) this is one of the world's largest extant species of skink. The extraordinary lizard has a blunt face like a tortoise, powerful feet like a monitor lizard and a prehensile tail like a possum.

An exclusive herbivore, it consumes a variety of plants but has a particular preference for the climbing aroids of the genus *Epipremnum*. Like giant reptilian caterpillars, Prehensile-tailed Skinks munch the soft foliage, leaving distinctive circular bite marks in the leaves. Those leaves contain high concentrations of irritating calcium oxalate crystals but the lizards suffer no ill effects.

The Prehensile-tailed Skink has a low birth rate, producing just one, rarely two very large young per litter. The developing embryo is nourished via a placenta, a rare phenomenon in reptiles. In order to cultivate the gut flora necessary for a vegetarian diet high in cellulose, newborn skinks consume the adult faeces.

Sadly many thousands of Prehensile-tailed Skinks have been poached from the wild to supply global pet markets. It has occurred at an alarming scale that eclipses the traditional use of these lizards as a food by local peoples. A mind-boggling 7,500 lizards were exported in just two years up until December 1990. Their export is now banned following a CITES Appendix II listing.

But with ongoing unsustainable logging it will be difficult or impossible for these slow-breeding vegetarian giants to regain their former numbers and populations are now severely fragmented. We were not just lucky to see a Prehensile-tailed Skink low down on that fig trunk. I fear we have been lucky to see one at all!

References: 34, 49

69

PACIFIC BLUE-TAILED SKINK

Emoia caeruleocauda

FAMILY:
Scincidae
LOCATION:
Haia Village, Chimbu Province,
Papua New Guinea
STATUS:
Least Concern, IUCN Red List

After a brief heavy downpour in a tropical rainforest in Papua New Guinea, water droplets gleam and sparkle on the saturated leaf litter. Suddenly there is a sinuous movement of eye-catching electric blue. Pacific Blue-tailed Skinks return to the sunny patches to resume basking on dead leaves and low vegetation. As these lizards move into the sunlight, waves ripple along their thin whip-like tails. It is as though the vibrant blue tails have lives of their own.

It seems counter intuitive for small edible lizards to draw attention to themselves but this is clearly what they are doing. It is highly likely the lizards are acutely aware of each other, so tail-waving could be an integral part of social structure and spacing. They may be akin to the arm-waving displays of some dragons and iguanas, directed at no-one in particular but everyone in general. But they are also obvious to predators.

When pursued, Pacific Blue-tailed Skinks are extremely alert, swift and evasive. It took me many attempts and a lot of stealth to get near them. From my experience of chasing other diurnal sun-loving lizards with brightly coloured tails, their conspicuousness seems proportional to their ability to evade capture. The colours are present on very quick lizards. If predators have missed these skinks previously and wasted energy in unsuccessful attempts, the message they receive may be: 'Don't bother. You know it is not worth trying.' A simultaneous counter message may be: 'Aim for the obvious part.' This diverts attention away from the head and body to the expendable tail, which can be regenerated.

Pacific Blue-tailed Skinks are common in New Guinea and have a vast distribution across the western Pacific, including many isolated islands that have never been connected to other landmasses. That is typical of the genus *Emoia*, with many of the nearly 80 species occupying far-flung islands. They have an ancestry of colonisation by rafting and drifting, and this includes a confusing array of other blue-tailed skinks.

Very similar colour combinations of black bodies marked with two or three prominent pale stripes and vivid blue or red tails have evolved among quite a few extremely swift skinks around the world. Sinuous tail waving is often a feature, and the colours are sometimes more pronounced in juveniles. Whether the target audiences are predators or their own kind, it seems that for some lizards, tails are communication tools. They have tales to tell.

References: 49, 98

70

BROWN SHEEN SKINK

Eugongylus rufescens

FAMILY:
Scincidae

LOCATION:
Wara Sera, Chimbu Province,
Papua New Guinea

STATUS:
Least Concern, IUCN Red List

For an impressively large, elegantly designed skink with a noticeable iridescent veneer, the Brown Sheen Skink does not often rate a mention among herpetologists. True, it does live in some remote locations in New Guinea, but it also occurs in far northern Cape York, a popular Australian destination for countless biologists, and on the islands of Torres Strait and the western Pacific. It is relatively common and people do find it without too much effort, so we ought to be more familiar with it.

Brown Sheen Skinks grow to a head-and-body length of about 15 centimetres (6 inches) with the tail slightly longer. It has an elongated body with short, well-clawed limbs and is quite powerfully built. Juveniles are boldly banded but adults from most areas lose pattern, retaining that beautiful iridescent sheen and sometimes barred lips. But the animals from some elevated altitudes in New Guinea keep their bands into adulthood.

This lizard pictured is one of those banded adults. It was accessed via a light plane flight from Goroka to a small airstrip followed by a two-day hike across rivers, over ridges and through limitless amounts of mud to reach a small research station set in dense rainforest. And there it was, hiding under a rotting board. Easy!

Brown Sheen Skinks like it dark and damp. All of the ones I have found were associated with rotting wood, typically in and under old logs at an advanced stage of breakdown. They occupy a variety of monsoon forests and rainforests, where they are typically terrestrial, but have also been recorded in tree hollows and tall root buttresses. On Rennell Island in the Solomon Islands they also shelter under limestone rocks.

Brown Sheen Skinks have a broad diet of invertebrates and lay small clutches of two to four soft-shelled eggs. It has been suggested that the small clutches derive from greater competition in the tropics, placing a premium on larger and more competitive offspring. The reason the young are so boldly banded is not understood.

For the most part the only published information on sheen skinks appears in field guides to Australia and parts of the Pacific, and in broader accounts of the herpetology in various regions. I think these large skinks with their banded offspring and populations of plain and patterned adults are overdue for some closer attention.

References: 30, 49

71

GREEN TREE SKINK

Lamprolepis smaragdina

FAMILY:
Scincidae
LOCATION:
Kerevat, East New Britain,
Papua New Guinea
STATUS:
Least Concern, IUCN Red List

There are large, bright green skinks perched head downwards on tree trunks on tropical islands through Wallacea to the western Pacific. They stand out boldly against the trunks and grow to a head-and-body length of about 10 centimetres (4 inches).

The first truly green lizard I ever saw in the wild was on a trunk 5 metres (16 feet) above me in the coastal town of Madang in Papua New Guinea. It was impossible to reach but tantalisingly close enough to observe in some detail. Fortunately Green Tree Skinks turned out to be common in the area.

Green Tree Skinks have a vast range over numerous isolated islands, where they are abundant in coastal and lowland areas including gardens and parks. Genetic evidence suggests their distribution is a result of random waif dispersal throughout their evolutionary history. They are almost exclusively arboreal, rarely descending to the ground, so it is likely that trees and vegetation mats uprooted and washed away during storms have provided the floating rafts for those unwilling passengers that were lucky enough to make landfall.

Their preferred trees are usually large, have bare trunks without too much epiphytic growth, and are typically isolated in semi-cleared areas and surrounded by low vegetation. In some areas, the trunks of coconut palms in plantations are popular. Green Tree Skinks rarely occur in closed-canopy forest, possibly because plenty of sun exposure on the trunks is an important requirement for these basking lizards.

Although they are considered to be exclusively diurnal, there are records of Green Tree Skinks living on trees growing close to lights feeding at night on attracted insects. They are generalist invertebrate predators that sometimes also capture other lizards such as small geckos, and occasionally eat ripe fruits.

Over most of their range Green Tree Skinks are simply bright green with dark-edged scales. However some colour variants, including those I saw in Papua New Guinea in the province of East New Britain were half green and half coppery brown, with the colours split across the middle. It was almost as though they could not decide what colour they really wanted to be!

References: 45, 49, 98

72

GREEN-BLOODED SKINK

Prasinohaema virens

FAMILY:

Scincidae

LOCATION:

Ngattokae Island,
Solomon Islands

STATUS:

Least Concern, IUCN Red List

I confess to being a bit disillusioned by the Green-blooded Skink. I was in the Solomon Islands to help a colleague find reptiles for research into blood parasites on tropical islands as part of a broader program investigating the invasive Brown Tree Snake (*Boiga irregularis*) on Guam. He was taking blood samples from a wide range of snakes and lizards, and looking over his shoulder I was expecting something more impressive from the famous blood of this unusual skink. But it looked red to me!

The bile pigment biliverdin is present in the plasma. There are higher concentrations in this lizard and its four close relatives than in any other organism. In fact 40 times greater than that of humans with green jaundice! As a result the bones, muscles, tongue and mouth-lining are a striking bluish green to lime green. Even the egg contents appear pale green through the shell.

This attractive green skink is a wholly arboreal forest dweller, tending to prefer thick vegetation and rarely if ever descending to the ground. It moves easily along slender stems, among foliage, and over smooth leaf surfaces, aided by a prehensile tail and by the unusual lamellae under its toes. These have microscopic setae which function like those of some geckos and anoles (see page 102), and in captivity Green-blooded Skinks have even been observed to climb the wall of a glass aquarium. At night they sleep exposed on the surfaces of leaves.

Green-blooded Skinks have a fixed clutch size of two eggs, laid in the decomposing leaf litter that accumulates in the forks of branches and among epiphytes.

I was certainly not disappointed to see Green-blooded Skinks navigating the vines, branches and foliage with great dexterity. However, when a single drop of blood was extracted from the base of a lizard's tail before it was released I had been hoping for something more dramatic. I naively misunderstood the complexities of blood and its components, and expected to see bright green fluid, like a kale smoothie.

References: 5, 31, 49

73

SCHMIDT'S CROCODILE SKINK

Tribolonotus schmidti

FAMILY:

Scincidae

LOCATION:

Tenaru, Guadalcanal,
Solomon Islands

STATUS:

Least Concern, IUCN Red List

Looking under damp rotting logs in a tropical rainforest is a pastime loaded with potential! For the lizards that live there, life in damp darkness poses a unique set of challenges. Most have very smooth, highly polished and close-fitting scales to repel moisture and allow easy passage, but some take a different approach.

Schmidt's Crocodile Skink has rough, strongly keeled, granular scales creating a somewhat crocodile-like texture. It is endemic to the island of Guadalcanal, and belongs to a genus of ten species that have radiated across the Solomon Islands and New Guinea. Some are highly distinctive in form but they share similar scale features.

The unusual texture of these lizards creates an increased surface area and it is believed this facilitates the dispersal of water over the skin so a droplet is not repelled but rapidly dissipates across the surface.

Similar unusual scale structures occur among unrelated skinks in rainforests of north-eastern Australia and South-East Asia, and in several other lizard families in the Neotropics. They are mostly associated with moist environments and many live a sedentary life under damp logs.

Crocodile skinks are rarely encountered active, and apparently spend much of their time within those sheltered domains. Several may be found together, often some juveniles with an adult female. They avoid direct sunlight and seldom venture far from cover.

I have peered under plenty of logs in my time. Yet curiously in distant rainforests around the world I have uncovered lizards of similar size, build and relatively slow speed with rough textures hiding in those dark, damp places. I had never seen crocodile skinks until I found them under those logs on Guadalcanal but it came as no surprise to see them. Nature, it seems, has a way of repeating itself.

References: 49, 95

74

GIBBER CTENOTUS

Ctenotus septenarius

FAMILY:

Scincidae

LOCATION:

South Galway Station,
Queensland, Australia

STATUS:

Least Concern, IUCN Red List

A disturbance in the leaf litter, a hint of shiny scales with a half-concealed striped pattern, then a blurred streak through the undergrowth. I try to at least get a glimpse to see what species it is, before the skink disappears into a concealed burrow somewhere under a tussock. Another elusive, nameless striped skink has performed a vanishing act!

In a continent of unparalleled lizard diversity, striped skinks of the genus *Ctenotus* are riding the crest of that success. It is Australia's largest reptile genus, with more than 100 named species occurring virtually everywhere except Tasmania. They like dry and well-drained habitats, largely avoiding wetlands and densely shaded areas. Spinifex deserts and seasonally dry tropics are the hot-spots, where large numbers of species coexist. Their impressive diversity is due mainly to their ability to partition resources.

Striped skinks range widely in size from head-and-body lengths of 4–12 centimetres (2–5 inches). This significantly affects their diets. Small species feed on various small invertebrates and some prefer termites. Larger ones eat a broad range of invertebrates and even smaller lizards including other striped skinks.

Species with proportionally shorter limbs lurk near the edges of vegetation while those with longer limbs forage much more widely across open terrain. Some are active earlier in the morning and others operate later at hotter temperatures.

Up to ten or more species can coexist, with competition reduced across at least three dimensions; where they live, what they eat and their activity times. But all are sharp-eyed and lightning fast.

Striped Skinks have been posing identification problems for years. Some have unique patterns of stripes or a combination of stripes and spots that are recognisable at a glance to those familiar with them. For others the differences are subtle, so a fleeting glimpse is not good enough in areas where many species occur together. Often the striped skink has gone before any field identification can be made. Even in the hand or laboratory there are still some arguments about what's what.

The Gibber Ctenotus has defined habitat preferences, occupying arid areas on exposed stony slopes and along sparsely vegetated gravelly drainage lines from southern Northern Territory to south-western Queensland. The colour and pattern is distinctive, comprising a striped forebody and rusty hindquarters. It forages extensively across bare open ground, and with those extremely long appendages it is one of the fastest among a genus of speedsters. Nearly all the ones I have seen were way too quick for me.

Reference: 59

75

CUNNINGHAM'S SKINK

Egernia cunninghami

FAMILY:

Scincidae

LOCATION:

Altona North, Victoria, Australia

STATUS:

Least Concern, IUCN Red List

A day out lying in the sun for mum, dad and the kids. And perhaps some aunties and uncles too. The adults might have a little salad if they feel like going out for it, while the youngsters may take a passing caterpillar or cockroach. Oh, and if anyone needs the loo, it is over there beside that rock. This is not exactly an action-packed day but the Cunningham's Skinks are sharing it as an extended family.

Complex social arrangements are rare among lizards, but some members of the genus *Egernia*, and particularly Cunningham's Skinks, are potentially the world's most social lizards. However, sociality is also well documented in other families including the Armadillo Lizard, (family Cordylidae) in South Africa (see page 138).

Depending on locality, and presumably the availability of shelter and resources, Cunningham's Skinks can live in communities of more than 20 individuals, although groups of about half a dozen closely related lizards are more common. These skinks can cohabit for several years, share shelter sites and often bask together with chins resting on backs and arms over shoulders.

There is safety in numbers, and many eyes are better than two. Cunningham's Skinks are able to detect predators earlier when in groups, compared to singly. One beats a hasty retreat and they are all gone simultaneously.

Colonies usually have a shared latrine site, with accumulated faeces covering up to 3 square metres (32 square feet). Scat piling may serve a social function, such as chemically marking a community as distinct from an adjacent one that is more distantly related.

Most populations of Cunningham's Skinks live among rocks, although some are associated with large fallen logs. They shelter communally in tight crevices and thanks to the modified, strongly keeled scales on their backs, raised to spines on their tails, they are extremely difficult to extract.

In keeping with other large-bodied skinks, Cunningham's Skinks are mainly herbivorous, with adults eating more than 90 per cent plant material. The diets of juveniles more closely resemble different species of adult skinks at a similar size, comprising mainly invertebrates such as insects. As they grow older and larger they drift towards eating more plants.

The joy of watching Cunningham's Skinks is that a whole family of assorted age groups can be observed together, basking in apparent harmony. The challenge is all those eyes, alert for any false move I make. If I spook one, they're all gone.

Reference: 16

76

MURRAY'S SKINK

Silvascincus murrayi

FAMILY:

Scincidae

LOCATION:

Mary Cairncross Park,
Queensland, Australia

STATUS:

Least Concern, IUCN Red List

As I hike the trails through the subtropical rainforests of mid-eastern Australia, I sometimes know I am being watched. Handsome glossy skinks with coppery-brown backs, yellow bellies and a fine dusting of bluish spots over their sides observe me with large liquid black eyes. They sit motionless on rotting wood beside the tracks, exuding an apparent air of confidence and trust, and showing little concern or inclination to flee as I walk past.

Murray's Skinks are widespread in rainforests of northern New South Wales and southern Queensland. They are particularly fond of large logs with cavities such as borer holes where they often rest with just the head and forebody protruding. Because of their confiding nature and willingness to perch beside walking tracks, they are often the most obvious reptiles to hikers in those lush habitats.

They are ambush hunters that scan the rainforest floor for invertebrates. These are usually insects but can also include worms and spiders such as the infamous funnel-webs that are lethal to humans. They are also occasionally reported feeding on fungi and fallen fruit.

The closest relative to Murray's Skink is Tryon's Skink (*Silvascincus tryoni*), which occupies a much more restricted range at higher elevations. It prefers the cooler climate prevailing among the ancient Antarctic Beech forests.

It has always been a fairly simple strata title arrangement. Murray's Skinks of the warmer low to mid altitudes gave way to the Tryon's Skinks at upper elevations where the temperatures are generally lower.

However, as I've climbed in the Border Ranges in recent years I have started seeing Murray's Skinks among the Beech Trees where their lofty cousins live. I wonder whether these two species have a long history of living cheek by jowl in this upland forest, or whether the dynamics of competition are shifting.

At this stage it remains a hunch, but I fear Murray's Skinks may, under pressure from rising temperatures, be drifting into new habitat options. I am also concerned Tryon's Skinks could forfeit the lowest parts of their habitats to warmer climates and lowland cousins, with no realistic options to move any higher. Driven by a changing climate it may be that Murray's Skinks have become like some human counterparts – the relations who arrived for a visit and just didn't leave!

77

GARDEN SKINK

Lampropholis delicata

FAMILY:

Scincidae

LOCATION:

Gerringong, New South Wales, Australia

STATUS:

Least Concern, IUCN Red List

They bask in my garden beds, forage on the exposed patio, and have turned up in every room inside the house. The versatile and aptly named Garden Skinks thrive in altered human environments, including suburbia where natural bushland is a distant memory.

Adaptable lizards enjoy human hospitality over most of Australia, with the species varying according to locality. In eastern Australia the Garden Skink is one of the most common lizards, thriving in millions of household gardens.

The benefit of having lizards around the home is simple. We can watch what they get up to. They may be just small brown skinks lacking flamboyant crests or vibrant colours of more spectacular lizards, but they are urban success stories and their lives unfold under our noses. From the comfort of people's homes I have been an observer of stand-offs between rivals, the stalking and capture of insect prey and the grasping of the female by the male while they mate.

Gardeners often uncover caches of Garden Skink eggs, concealed under bricks and other humid sheltered sites. Such aggregations, numbering up to one hundred or more, represent the combined efforts of numerous female skinks, sometimes of several species. Many of these would have travelled from neighbouring properties, to lay communally. The presence of old decayed eggs from previous seasons indicates repeated use.

Garden Skinks are not obliged to nest communally, and single clutches are often encountered. But the frequency of large caches indicates that it is common behaviour. This 'all eggs in one basket' approach has been adopted by more than 200 lizard species worldwide. Benefits may include diluting predation on any one particular egg clutch, and ongoing use of sites with a proven record of successful hatchings. The obvious downside is the potential for a predator to scoff the lot, wiping out the collective annual reproductive output of most females over an area.

Humans have unwittingly dispersed Garden Skinks to some areas where they are not welcome. New Zealand's endemic and threatened skinks must now deal with introduced and competitive Garden Skinks, known to New Zealanders as Plague Skinks! They have also

reached Lord Howe Island. The local Lord Howe Island Skink (*Oligosoma lichenigera*), a Vulnerable species, lives in exile on small offshore islands due to historical rat invasions. Its potential for reintroduction following successful rat removal will not be helped by an abundant new introduced lizard on the scene. They are even on Hawaii too. At least they belong in my garden, where they are both welcome and entertaining.

References: 88, 95

78

KEELED SLIDER

Lerista planiventralis

FAMILY:
Scincidae
LOCATION:
Point Quobba, Western Australia
STATUS:
Least Concern, IUCN Red List

Crazy meandering tracks create complex looping designs on the sand. So many tangled paths that it is difficult to single one out and plot its course. There! I see where one ends, bend down and scoop up a handful of loose sand. Blowing it out of my cupped hands I am left holding a Keeled Slider.

The world of a Keeled Slider consists of that upper few centimetres of loose sand. That is where it essentially lives its entire life. For this superbly designed sand-swimmer, fine dry sand is a medium that is as fluid as water.

This is one of nearly 100 species of *Lerista*, an Australian genus that collectively exhibits a near-complete range of limb and digit reduction. They range from relatively unmodified surface-active leaf-litter inhabitants with well-developed pentadactyl limbs, through stages of limb and digit reduction and loss to completely limbless species that burrow in loose substrates.

With just two fingers and three toes, the Keeled Slider sits about halfway along the scale in terms of limb reduction, but it is one of the most highly modified for locomotion in loose sand.

Its acutely shovel-shaped snout easily penetrates the sand while strong longitudinal keels along either side of the belly provide traction for forward motion, propelling it as it wriggles to create undulating tracks. Its limbs are short and it has lost some digits but the three toes on each foot remain relatively long. They may help push it through the sand.

Keeled Sliders are active by day, moving easily just beneath the exposed surface as they hunt small invertebrates. Every so often a head pops up like a miniature monster in that sandy sea, then vanishes and the winding tracks proceed forward.

As I walk, I admire the fancy patterns being traced at my feet. Honing my skills I scoop up another couple of sliders and release them. But then I have something different in my hand, a tiny yellow-and-black ringed snake. The small, weakly venomous West Coast Banded Snake (*Simoselaps littoralis*) is also a skilled sand swimmer with a protruding flattened snout. The upper sand layer is its hunting ground, where it preys exclusively on small lizards, particularly sliders!

These are more than just attractive wavy lines in the sand. They are the imprints of the life-and-death dramas of the hunters and the hunted unfolding a centimetre or two under the surface.

79

SANDSTONE WORM-SKINK

Praeteropus auxilliger

FAMILY:

Scincidae

LOCATION:

Lake Elphinstone,
Queensland, Australia

STATUS:

Not evaluated but recommended as
Endangered, IUCN Red List

On the sandstone outcrops in a very restricted area of mid-eastern Queensland, a loose compost of sandy soil mixed with decomposed plant debris accumulates in fissures between the boulders. Shielded under layers of newer leaf litter these friable humid pockets are the domain of a newly named species of completely limbless skink.

Loss of limbs has occurred among more than 100 species of skinks in three major lineages around the world. This means the process has evolved independently many times. Wherever skinks have access to a soft medium suitable for burrowing, the pressure is there to exploit it in a variety of ways, including wriggling right in.

To progress easily through the compost, the Sandstone Worm-Skink has undergone an extreme makeover. The body has become elongated and worm-like, eyes are greatly reduced, ears are concealed under scales, head is slightly flattened, and a thickened cover has formed to protect the scales on the snout and chin. And it has lost all external traces of limbs.

The process of moving through a yielding substrate as opposed to digging tunnels is called 'substrate-swimming'. Those that penetrate loose sand (sand-swimmers and sand-divers) often have streamlined, finely chiselled snouts for rapid, fluid progress (see pages 164 and 176). But for a moister and more cohesive medium like organic compost, a round, armour-plated snout pushed forward by a cylindrical worm-like body and thick blunt tail is an effective way to proceed.

The Sandstone Worm-Skink has limbless close relatives in other parts of Queensland, but these occur in at least some protected habitats and some occupy large distributions. The Sandstone Worm-Skink's range is highly restricted, with the two known populations only about 35 kilometres (22 miles) apart. Neither of these sites is protected, and it is unlikely any populations will be found in protected areas. The intervening region has undergone extensive land clearing, grazing and weed incursion. For these reasons, it is proposed that this highly modified burrowing species, named as new to science in 2021, qualifies as Endangered.

Reference: 40

80

PYGMY BLUE-TONGUE

Tiliqua adelaidensis

FAMILY:

Scincidae

LOCATION:

Burra area, South Australia

STATUS:

Endangered, IUCN Red List

I still reflect on the words I published in 1988: 'Possibly extinct.' Happily they were proven wrong thanks to a case of pure serendipity, but it was another 20 years until I saw a Pygmy Blue-tongue in the field.

In 1959 a labourer at Marion near Adelaide poured hot tea down a hole. Two traumatised lizards ran out. There would be no more records of the Pygmy Blue-tongue for 33 years despite decades of targeted searches in remnant patches of bushland.

The Pygmy Blue-tongue became the Holy Grail of Australian herpetology, and everyone visiting the Adelaide area dreamed of rediscovering it. Reluctantly it was conceded that this may be the first extinction of an Australian reptile since European occupation.

The species was suggested to have occurred in mallee communities, perhaps hiding in stump hollows and under limestone. Insightfully, based on the unusual features of preserved specimens, it was proposed that the Pygmy Blue-tongue had been an ambush predator that hid in hard-rimmed hole refuges and was able to turn in narrow spaces.

Fast forward to 1992. Two herpetologists collected a road-killed Eastern Brown Snake (*Pseudonaja textilis*) in treeless grassland near Burra in South Australia. They opened it to examine a lump in the belly and encountered the intact remains of a Pygmy Blue-tongue. This was dead proof that the species was still alive!

Small wonder it was overlooked. No-one had searched in treeless grazing country, and certainly no-one had been peering down spider holes. These, it turns out, are the hard, tight spaces it was predicted to occupy.

The life of each lizard is centred on the vertical shaft of a mygalomorph spider burrow. It must locate one that no longer contains a spider! From there it can ambush passing invertebrates (also insightfully predicted) and block the passage with its large rugose head, yet its soft flexible body allows it to turn within the confined space.

Pygmy Blue-tongues are Endangered. They can occur in high densities, but subpopulations are fragmented, dispersal is limited or impossible, and none exist in significant conservation areas. Some subpopulations have even disappeared while they were being monitored. The lizards do well with traditional sheep grazing but conversion to cropland, pesticide uses and any changes to current farm practices elevate the threats.

It was exhilarating to walk in those undulating grassy fields, peer down insignificant-looking holes with a laparoscope and see the face of a Pygmy Blue-tongue looking up at me. Like Lazarus, resurrected from the dead, decades after I and other authors wrote of its possible extinction.

References: 17, 29, 96

81

EASTERN BALKANS GREEN LIZARD

Lacerta diplochondrodes

FAMILY:

Lacertidae

LOCATION:

Apolakkia, Rhodes, Greece

STATUS:

Not listed but probably qualifies as
Least Concern, IUCN Red List

There is a tiny stone church in the west of the Greek island of Rhodes. The medieval temple looks like it was haphazardly thrown together, but the walls are lined with 13th-century frescos. Awed by the magnificence, I stepped blinking out of the dim light into the bright Mediterranean sunshine and noticed a large, bright green lizard running along the edge of the exterior stonework.

There are about 10 species of green lizards distributed across Europe and Russia. They are a characteristic and spectacular component of European fauna, but have been a confusing group with plenty of local variation and many described subspecies.

The Eastern Balkans Green Lizard was only recognised as a distinct species in 2019. We generally equate new species discoveries with expeditions to remote places, but not hiding in plain sight in the cradle of Western civilisation! Its 'discovery' resulted from a close examination of the races within a widespread and variable species.

Research has shown that the subspecies on islands from the eastern Aegean Sea to Turkey, Bulgaria and Romania, including the population on Rhodes, is sufficiently distinct to warrant elevation to the rank of full species. And with a head-and-body length of 16 centimetres (6 inches) it is one of the largest lizards over much of that range.

Green lizards are lacertids, the dominant lizard group in Europe. In many respects, as surface-active sun-loving lizards, they fill the niches occupied by some skinks over much of the world, and by teiids in the Americas. For Europeans, lacertids are the 'typical' lizards that bask along the sunny edges of paths or among vegetation, and are common in gardens.

There are some highly modified lacertids in Africa and Asia, but European lacertids are what we consider as conservatively 'lizard-shaped,' having four limbs each with five digits, rounded snouts, smooth cylindrical bodies, moderately long tails and no crests, dewlaps, flaps or flanges. Indeed *Lacerta* is Latin for 'lizard' and forms the root of Lacertilia, which is the term we use for all lizards.

Eastern Balkans Green Lizards can now hold their colourful heads high with recognition as a species in their own right. And like so many of Europe's lizards, they add a splash of colour and charm to parks, roadside thickets and the medieval and classical antiquities where they continue to thrive.

Reference: 43

82

MILOS WALL LIZARD

Podarcis milensis

FAMILY:

Lacertidae

LOCATION:

Milos, Greece

STATUS:

Vulnerable, IUCN Red List

In the Greek islands, the sky and the sea really are as blue as they appear in the postcards. As I travelled between islands, those deep hues were a constant part of the region's picturesque charm. Less constant were the lizards on different islands. Some were widespread species extending from the Greek and Turkish mainlands and spanning numerous islands, but there were also narrow endemics restricted to small island groups.

On Milos in the Cyclades group, Milos Wall Lizards sun themselves and hunt among the olive trees, on dry-stone walls, in home gardens, on the marble ruins of past civilisations and in local dry shrublands. Females are brown, often with well-defined pale stripes. Males are distinctive and extremely attractive, having black flanks with large green, blue or turquoise spots.

These lizards are so abundant it is easy to forget how restricted they are. Milos Wall Lizards live only on a small group of islands; Milos, Kimolos, Polyaigos and Antimilos, and the rocky islets of Ananes, Falkonera and Velopoula. Even on that mere cluster of specks in the blue Aegean Sea, populations of Milos Wall Lizards have diversified into three separate subspecies with differences in pattern. Clearly there is little opportunity for dispersal between landmasses.

Like all small lacertids, Milos Wall Lizards feed mainly on arthropods, but they differ in what kinds they eat. During summer there is a stronger reliance on ants – something that many other small lizards tend to reject.

They share their isolated habitat with a rare endemic snake, the Cyclades Blunt-nosed Viper (*Macrovipera lebetina schweizeri*), and the wall lizards form a significant of part its diet, particularly for young snakes.

I spent many hours patrolling the edges of stone walls to observe the lizards. I also put in an extraordinary amount of effort to see a viper. There were plenty of lizards and I eventually located a snake, but some would argue the time I spent indulging in the world of Greek reptiles was excessive. I may not agree but I can see the merit in that argument. After all I was on my honeymoon!

References: 1, 86

83

ITALIAN WALL LIZARD

Podarcis siculus campestris

FAMILY:
Lacertidae
LOCATION:
Figline, Italy
STATUS:
Least Concern, IUCN Red List

As I watched lizards basking beside stone steps I wondered just how many others before them had taken up those same sunny positions. And I mused on the number of human feet that must have walked past those lizards. More than two thousand years of pedestrian traffic, from the sandalled feet of robed senators, noblemen, merchants and thieves, to today's tourists and the touts who try to fleece them. Italy's awe-inspiring ruins are ideal habitats for Italian Wall Lizards. I saw their heads protruding from cracks in the stone walls of Pompeii, the Roman Palatine, the Colosseum and the amphitheatre of Fiesole. They basked and they watched the feet go by.

The very name, Italian Wall Lizard, speaks volumes about an ability to thrive in the modified habitats created by humans. And its abundance in famous archaeological sites has led to an alternate name, Ruin Lizard. When it comes to human impact on wild animals, this is one of the winners, cited under the IUCN Red List as 'increasing'.

The species occurs naturally throughout Italy, where it is one of the most common lizards, and its range extends to neighbouring countries. There are also introductions of invasive populations into Spain, Turkey, the UK and North America. A staggering 60 or more subspecies have been described – a testament to its variability across a range of habitats.

One explanation for this immense diversity within one species may lie in its ability to evolve rapidly when faced with new or different environments. After 36 years – a blink in evolutionary time – a population of Italian Wall Lizards introduced to an island off Croatia was found to have measurably diverged in the shape and size of the head, strength of bite force and even the development of new structures in the lizards' digestive tracts to accommodate a shift in diet. There were also behavioural changes in population density and social structure.

Italian Wall Lizards feed mainly on arthropods, but perhaps part of their strategy for success lies in the diversity of diet. They eat more plant material than other similarly sized European lacertid lizards, will capture other small lizards, and have even been observed consuming carrion in the form of a dead shrew.

Italian Wall Lizards are thriving today on contemporary structures as well as ancient ruins. If any of our modern buildings still stand in two thousand years I wonder if there will be lizards peering from the cracks. And will there still be human feet to walk past them?

Reference: 37

84

SHOVEL-SNOUTED LIZARD

Meroles anchietae

FAMILY:
Lacertidae
LOCATION:
Swakopmund area, Namibia
STATUS:
Least Concern, IUCN Red List

A little lizard darts along the sloping face of a large sand dune in the Namib Desert, generating miniature avalanches of loose sand with each step. I wanted to get closer but sneaking up is pointless. There are no trees or rocks, almost no bushes or grasses. Just unstable slopes of sand, shifting with the vagaries of the wind. And of course me, the sole prominent feature in a sandy realm under the watchful eye of an alert Shovel-snouted Lizard.

The first couple just sprinted over the crest of the dune, never to be seen again. But if I ran towards a lizard, it dived forwards, plunged head-first into the sand and vanished. The trick was to see exactly where it disappeared, leaving only a blemish of disturbance on the surface. Then it could be scooped up in a handful of sand.

The Shovel-snouted Lizard is a sand-diver, a swift, active surface-forager that takes refuge in loose sand by diving in as if the medium was fluid. This specialised behaviour requires extreme modifications such as a broad, acutely shovel-shaped snout extending well beyond the lower jaw, well-developed limbs and long fringes of slender scales under the toes. Sand-diving is uncommon but those unusual features and behaviour have evolved convergently in several unrelated lizard families in sand deserts of Africa and North America, and some sand-swimming skinks have a similar snout design (see page 164).

The crazy antics of Shovel-snouted Lizards are quite spectacular. They dance a weird jig, alternately raising two diagonally opposite feet at a time. This has been termed a 'thermal dance', to avoid cooking their tootsies on the hot sand, and that may be the case in hot weather. However conditions were mild when I watched the performance, and the choreography even included a nifty move when all feet were simultaneously held aloft as the lizard rested on its belly. So there is probably more to these odd capers than just thermal management.

Shovel-snouted Lizards are endemic to unstable dunes of the Namib Desert. It is a harsh environment of shifting sand and high temperatures, and an ecology dominated by aeolian forces. Most small, active desert lizards are largely or exclusively arthropod feeders but Shovel-snouted Lizards are omnivorous. In a land largely devoid of vegetation, winds deliver much of their sustenance in the form of insects and even grass seeds, which constitute a large proportion of their diets. Because they rarely if ever access surface water, all moisture is obtained from their food and condensation from fog. Those little dancing lizards live in a tough regime!

References: 15, 54, 68

85

WEDGE-SNOUTED LIZARD

Meroles cuneirostris

FAMILY:

Lacertidae

LOCATION:

Sesriem area, Namibia

STATUS:

Least Concern, IUCN Red List

In Namib-Naukluft National Park, red desert dunes rise abruptly from expansive, open gravel plains. The transition between two starkly different habitats is as sharply delineated as, well, a line in the sand.

The bases of those dunes, where the loose sand is sparsely vegetated with shrubs and tussocks, are the domain of the Wedge-snouted Lizard, a racy lacertid with a rusty colour and flecked pattern that closely matches the substrate.

Wedge-snouted Lizards deal with potential danger using evasion and speed. I saw them avoiding me on those dune slopes, red streaks that vanished into the distance propelled by long legs. It has been found that the evasive actions taken by juveniles differ from those of adults. While larger lizards often dash into the cover of one of the meagre clumps of vegetation, there is a greater frequency for smaller ones to skirt the shrubs and tussocks. For those young lizards, the dappled shelter of vegetation may harbour greater threats than the object they are fleeing. Those dangers include lurking vipers, predatory skinks and even cannibalistic adult Wedge-snouted Lizards!

For the most part Wedge-snouted Lizards are arthropod feeders, although separate studies have identified different insect groups as comprising the bulk of the diets. Some show a bias towards caterpillars, others to beetles, wasps, ants or termites. For opportunistic insectivores, the differences probably reflect what was most available at the times and places of each survey.

Wedge-snouted Lizards are also occasional thieves. They have been observed lying under the shade of bushes monitoring the ant trails. Unburdened ants are allowed to pass unhindered but those carrying food are robbed of their cargo. Such acts of piracy are considered forms of kleptoparasitism.

References: 28, 54, 69

86

SIX-STRIPED LONG-TAILED GRASS LIZARD

Takydromus sexlineatus

FAMILY:
Lacertidae

LOCATION:
Bogor, West Java, Indonesia

STATUS:
Least Concern, IUCN Red List

One of the most interesting things I saw at the Jakarta Zoo was actually in a garden bed rather than an enclosure. A soft blurred movement among low ornamental grasses and ferns turned out to be a Six-striped Long-tailed Grass Lizard.

This extraordinarily slender lizard is superbly adapted to disperse its light weight across low fine foliage. That includes a thin tail up to five-and-a-half times the length of its body, making it one of the longest tails, relative to head-and-body length, of any extant lizard.

Six-striped Long-tailed Grass Lizards are traditionally grassland specialists, where they can move and bask on the ends of long blades of grass. Effectively they are 'grass-swimmers' and leapers, projecting themselves forward with a flick of that attenuate tail and able to slip through the unstable foliage with ease, speed and discretion.

They thrive in disturbed areas too, including grassy roadside verges, fields and gardens with shrubs and ferns. A homeowner in Bogor had them in his garden, including one which he considered a 'regular'. He told me it lived among his ferns where it would often rest on fronds up to 1 metre (3 feet) above the ground.

The Six-striped Long-tailed Grass Lizard belongs to a family of lizards common throughout Europe and Africa that are generally conservative in their physical structure. However this species and several others in its genus have evolved extreme morphology to exploit a specialist component of many habitats, the lightweight outer foliage of grasses, shrubs and fern tips. It is also one of the few lacertids to penetrate South-East Asia.

I have been fortunate to see this strange lizard at several localities, including four in the Jakarta Zoo garden beds that day. Those glimpses lasted for just the few seconds needed to register what they were, before their slender forms vanished. I never saw any of them again.

Reference: 33

87

WESTERN WHIPTAIL

Aspidoscelis tigris

FAMILY:

Teiidae

LOCATION:

Ajo, Arizona, United States

STATUS:

Least Concern, IUCN Red List

A small lizard dashes between two shrubs on an exposed gravel plain. With a keen eye and some flicks of its forked tongue it quickly checks for invertebrates, and scuttles rapidly to the next shrub. It is furtive and alert, ready to vanish into a burrow or crack at the slightest sign of danger. The lizard is a Western Whiptail.

More than 20 species of whiptails or racerunners occur across the United States, where they are among the most common diurnal lizards in open, often sparsely vegetated habitats. But getting near enough for a better look is not easy. They appear nervous, foraging with rapid jerky movements, and vanish at any hint of disturbance.

A number of different families around the world play a role as small to medium-sized diurnal lizards that move through the terrain, hunting by actively foraging and investigating, rather than perching and watching. Most of those fast little lizards in the United States and Mexico are teiids. Typically, teiids have granular body scales and large rectangular plate-like scales across the belly.

Some species of whiptails in the United States exist as all-female populations. There are no males. When they mature, they reproduce clones of themselves, and experiments using skin grafts applied successfully on lizards collected hundreds of kilometres apart have shown entire species comprising individuals that are genetically identical.

Identifying the different whiptails can be confusing. Some are extremely similar and several change in appearance as they grow. The Western Whiptail is one of the most variable. Over its broad distribution covering much of the western United States and Mexico, a confusing array of up to 15 subspecies are recognised – or not recognised, depending on what taxonomy is followed. On a colour-coded distribution map those variants look like a multihued patchwork!

Western Whiptail ecology varies across its broad geographic range, which is not surprising given the number of described races. Lizards in the north are active for shorter periods, at cooler ambient temperatures and with lower core temperatures. There is also a lower frequency of broken and regrown tails in the north, suggesting their southern cousins may face a greater or more innovative range of predators.

The lizard pictured, from an arid area of southern Arizona just a few kilometres from the Mexican border, still has its original tail. It has been lucky so far.

References: 18, 60

88

NORTHERN CAIMAN LIZARD

Dracaena guianensis

FAMILY:
Teiidae

LOCATION:
Añangu Creek, Rio Napo,
Amazon Basin, Ecuador

STATUS:
Least Concern, IUCN Red List

With smooth crocodile-like skin and an immense swollen head, the lizard was an impressive sight to behold lying full length, its tail hanging down, along a branch above the water.

I was drifting silently along a narrow tributary of the Rio Napo in Ecuador's Amazon Basin, in still water edged by tropical rainforest. The silence was punctuated by the noise of squirrel monkeys cavorting overhead. It was an idyllic scenario to encounter a Northern Caiman Lizard.

The lizard was not remotely disturbed by my approach as I sat at the front of a canoe, telephoto lens in hand. It was so still it could have been a model. In fact, when I returned a few hours later it was still in the same place.

At nearly 1 metre (3 feet) long, the Northern Caiman Lizard is among the larger lizard species in the Amazon Basin. It enjoys a vast distribution across tropical northern South America, along drainage systems from Peru and Ecuador to eastern Brazil. It is often seen lying on riverbanks or, like my lizard, on branches over water. Amphibious and a graceful swimmer, it propels itself forward using its laterally compressed tail, with the head up and limbs pressed to the sides. It will readily dive in if disturbed.

The Northern Caiman Lizard has the most specialised diet of the large teiid lizards, feeding almost entirely on molluscs during the wet season. That is where an immense head, strong jaws and large molar-like rear teeth come in handy for crushing the shells. They swallow the contents and discard the fragments. Northern Caiman Lizards frequently enter flooded forest to forage, and in shallow water will sometimes walk along the bottom in search of prey. When conditions are dry they often hunt in trees for insects and possibly to raid birds' nests.

The sexes of adult Northern Caiman Lizards can be distinguished by head size. Those of males are enormous and appear disproportionate to body size. They have become so large, and their jaw muscles so huge, one wonders how they do not hinder normal functions.

When I left that tropical paradise a week later, my Northern Caiman Lizard was still pretty much in the same place. However, it had changed branches and was facing in a different direction. So it was definitely a live lizard and not a model.

References: 8, 62

89

DELICATE WHIPTAIL

Holcosus leptophrys

FAMILY:

Teiidae

LOCATION:

Corcovado National Park, Costa Rica

STATUS:

Least Concern, IUCN Red List

In the forests of Central and South America a sudden rustle among the leaf litter is usually a whiptail. Tropical whiptails are often heard before they are seen, because they usually notice the observer first, and disturb the leaf litter as they take evasive action.

On the rainforest floor, assuming it's a clear day, shafts of sunshine create puddles of sunlit leaf litter. These illuminated warmer patches fade as clouds move over, and shift as the sun crosses above a complex canopy. This mosaic of sunlight and shade is the domain of whiptails. They are also associated with gaps in the canopy where a tree has fallen, and along forest edges beside roads, tracks and watercourses.

Whiptails are alert, sun-loving, surface-active foragers that regulate their temperatures by basking and shifting between sun and shade. They are wary and constantly poised to flee if they perceive danger.

Many teiids prefer dry open habitats but Delicate Whiptails are common in the rainforest at Corcovado National Park on Costa Rica's Pacific Coast. They also often venture deeper into the shadier areas of the forest than other whiptails.

After an initial disturbance that rustles the leaves, it takes only a minute or two of patient waiting before the lizards perceive the threat has passed and resume their hunting. They move methodically through the leaf litter, checking in cavities and investigating under logs, always ready to seize any invertebrate that betrays itself with a wrong move.

As well as visual cues, whiptails are 'taste' feeders that identify food using chemoreception. With constantly flickering, deeply forked tongues they sample possible food items, transferring chemical data to the brain for rapid analysis via the Jacobson's organ in the roof of the mouth.

Watching lizards actively engaged in their daily routine, paying no heed to the observer, is an addictive pastime. My challenge when walking in dappled sunlight along those forest trails, was to try and see a whiptail before it saw me. In the end I figured the score was about even, but that doesn't include the unknown number that watched me unseen as I passed by.

References: 44, 71

90

SLOW-WORM

Anguis fragilis

FAMILY:
Anguidae

LOCATION:
Corfe Bluff, Dorset, United Kingdom

STATUS:
Least Concern, IUCN Red List

In one of Enid Blyton's books, a child sleuth in 'The Famous Five' crime-busting gang kept a Slow-worm in his blazer pocket, where it gave birth to a litter of shiny, silvery young. Even as a kid reading the book I wondered whether the Slow-worm was okay with that. There was also a pet Slow-worm named Sally Slither in Blyton's *The Mountain of Adventure*. The presence of Slow-worms in iconic British literature indicates how well-known this small limbless, semi-fossorial lizard is among the general populace, although it does have a long history of being mistaken for a snake.

Growing up to 40 centimetres (16 inches), the Slow-worm is the longest of Britain's native lizards, but there are only three, so it is not rising above a large pack! It is found over virtually all of Britain and across most of Western Europe, although it is absent from Ireland. The Slow-worm can lay claim as the only reptile in the Outer Hebrides, too.

Slow-worms inhabit heaths, commons, woodlands, parks and gardens, where they hide by day under stones and logs or in loose soil. Although gravid females frequently bask, Slow-worms generally avoid direct sunlight, emerging at dusk and at night to hunt invertebrates. Slugs are particular favourites.

They hibernate in winter, sometimes several metres below ground. Occasionally they aggregate in 'balls' of entwined Slow-worms, numbering more than one hundred.

If grasped, Slow-worms can easily lose their tails and grow new ones, a feature common to many lizards around the world. This ability to 'break into pieces' may have influenced Carl Linnaeus, the botanist and taxonomist who formalised our modern system of classification, to coin the name '*fragilis*' when he described the species in 1758.

For a lizard of its size, the humble Slow-worm may hold a longevity record. A Danish museum maintained a captive specimen for 54 years. It would actually have been older because its age at the time of acquisition was not known. Perhaps its name could have been Methuselah!

There were very few reptile books about Australia when I grew up during the 1950s and 1960s. My herpetological literature was biased towards the Northern Hemisphere, and so was the fiction such as Enid Blyton. So when I found Slow-worms under boards on a Dorset heath in 2009 it was almost like meeting old friends. And who knows, given how long they live, some of those adults could actually date back to my early reading days.

Reference: 90

91

MADREAN ALLIGATOR LIZARD

Elgaria kingii

FAMILY:

Anguidae

LOCATION:

Portal, Chiricahua Mountains, Arizona, United States

STATUS:

Least Concern, IUCN Red List

A Madrean Alligator Lizard pushes through the leaf litter, sliding over the pine needles to find a discreet sunny patch to bask. It is sleek and shiny, relatively short-limbed, long-tailed and slender, yet powerfully built. Its scaly armour is arranged as transverse rectangular plates, and a fold of soft skin covered with small bead-like scales runs along its lower flanks. That gives it the flexibility to twist and turn, and the capacity for expansion to breathe, feed and develop eggs. The texture of its back has a vaguely crocodilian look. That is why it is called an alligator lizard.

This Madrean Alligator Lizard was foraging in woodland near the town of Portal in the Chiricahua Mountains in south-eastern Arizona. That cool, elevated enclave is one of the 'Madrean Sky Islands' where lofty pine and oak woodlands are biodiversity hot-spots that are isolated in mild climates surrounded by warmer, drier desert lowlands.

Madrean Alligator Lizard is named after those sky islands of Arizona, New Mexico and adjacent Mexico. It is primarily a mountain dweller occurring at elevations from about 730 metres (2,400 feet) to more than 2,700 metres (8,860 feet) above sea level, where it occupies a range of habitats including creek edges, canyons and valleys, typically in moist areas. It likes plenty of leaf litter and other loose ground cover, where its long smooth body allows easy passage and the short limbs can propel it forward through twigs and other debris but not get in the way. It is usually heard rustling in the leaves before it is seen.

This is one of seven species of alligator lizards in the genus *Elgaria* found in the United States and Mexico. The Madrean Alligator Lizard is diurnal to nocturnal according to weather, and it hibernates during winter. It can respond vigorously if grasped, writhing, biting and defecating on its captor. It can also readily discard its tail and grow a new one.

It belongs to the diverse family of lizards called anguids, whose disparate members are so variable that some of them bear little resemblance to each other. Anguids range from completely limbless worm-like species that live in soil and leaf litter like the Slow-worm (*Anguis fragilis*) of Britain and Europe (see page 188) to arboreal members with well-developed limbs. Most of their diagnostic features are internal, relating to teeth and skulls. They all have hard scales containing bony internal structures called osteoderms, and like the alligator lizard, the fold of soft, finely scaled skin on the flanks is typical for most members.

Reference: 81

92

HIGHLAND ALLIGATOR LIZARD

Mesaspis monticola

FAMILY:

Anguidae

LOCATION:

Cerro de la Muerte, Costa Rica

STATUS:

Least Concern, IUCN Red List

It gets cold on Cerro de la Muerte. Freezing cold! Rising high above Costa Rica's hot tropical lowlands, the 'Mountains of the Dead' are named for the hapless travellers who perished trying to cross the range. They were ill-prepared for the extreme conditions and froze to death.

I travelled there in the relative comfort of a bus, although the winding road with sickening drops off the edge and the 'flexible' road rules made for a high-risk journey that still haunts me.

In that tropical alpine zone, brisk temperatures and páramo vegetation comprising low shrubs and grasses were a far cry from the rainforests I had left behind. I had entered another world. By mid-morning there were still white patches of frost clinging to the shaded sides of rocks and tussocks. But on the sunny side there were lizards basking. I was at 3,100 metres (10,200 feet) above sea level.

Although very common, Highland Alligator Lizards are among the few lizard species that live at these altitudes, occurring from 1,800–3,800 metres (5,900–12,500 feet) above sea level. Like many other lizards in cold climates the world over, they have the dark colours and stocky build needed to absorb heat and retain it. They make use of the sun when it is available, but when it is rainy, cloudy or their domain is enveloped in fog (which is very often!) Highland Alligator Lizards remain inactive under the cover of surface objects such as rocks.

Also in keeping with many cold-adapted reptiles, these lizards are live-bearers. Caches of eggs are not viable in consistently low temperatures but developing embryos benefit from the thermoregulatory behaviour of the mother. They give birth to two to ten young, which have been observed to remain in close association with their mother for period of time.

My visit to the chilly heights of Cerro de la Muerte was easy. A few hours on a bus along the Pan-American Highway, comfortable lodgings with close access to lizards in the surrounding shrublands, then a return journey by bus. It is a far cry from the hazardous crossings on foot and by horse that took many days and cost lives. I'm sure lizards were the last thing on the minds of those travellers.

References: 44, 71

93

GILA MONSTER

Heloderma suspectum

FAMILY:
Helodermatidae
LOCATION:
Tortolita Mountains,
Arizona, United States
STATUS:
Near Threatened, IUCN Red List

The Gila Monster is probably the most legendary lizard in the United States. The evocative name suggests something both extraordinary and menacing. It has achieved world notoriety because it is a venomous lizard. But like many creatures perceived as dangerous, fiction largely overrides facts.

The family Helodermatidae contains five species, including the beaded lizards, distributed from the south-western United States to Guatemala. The Gila Monster inhabits the deserts of the southern United States and Mexico.

Family members are almost exclusively raiders of vertebrate nests, eating litters of young mammals, reptile eggs, and birds' nestlings and eggs. This requires extraordinary chemosensory capacity, considerable time spent searching for patchy and well-concealed resources, and an ability to endure extended lean periods.

Gila Monsters have deeply forked tongues receptive to minute traces of organic particles, and an exceptional ability to determine the direction of a scent source. It has been suggested that the tongue, broader and flatter than the forked tongues of monitors and snakes, doubles up as an 'egg spoon' to scoop up the contents of broken eggs.

Gila Monsters have venom glands in the lower jaws, visible as bulges below the lips. Birds and mammals are extremely susceptible, reptiles less so and invertebrates are largely immune. The grooved teeth are not structured to inject venom but to deliver it quickly into the site of a bite.

The excruciatingly painful bite can produce extremely serious and unpleasant symptoms, but reports of human envenomation contain more hearsay than facts. History is littered with lurid accounts of death, misery and disfigurement from Gila Monster bites but they are associated with sketchy details, poor health and fertile imaginations.

The 17 reported bites by Gila Monsters and beaded lizards based on first-hand accounts or peer-reviewed literature since the 1950s all relate to pet lizards, animals being captured and mishaps during demonstrations. None involved accidental encounters. The last recorded human fatality from a Gila Monster was a drunk pool-hall operator bitten on the thumb while harassing one in 1930. It is generally accepted that no person in good health who has received appropriate medical attention has ever died from a Gila Monster bite.

Medical research reveals Gila Monster venom as a complex cocktail of active ingredients with extraordinary properties. One of the many components, helodermin, binds to breast-cancer cells and inhibits growth and multiplication of lung-cancer cells. It is unique to the venom of Gila Monsters. Other compounds have yielded promising results in the treatment of diabetes.

A person happily minding their own business has an extremely slim chance of coming to any harm from a Gila Monster, maligned victim of more than a century of misinformation and half-truths.

Reference: 12

94

KOMODO DRAGON

Varanus komodoensis

FAMILY:
Varanidae
LOCATION:
Rinca Island, Indonesia
STATUS:
Endangered, IUCN Red List

There is never a convenient place to fall and break a leg. That would certainly include the Indonesian islands of Komodo and Rinca, home to Komodo Dragons, the world's largest surviving lizards. They specialise in attacking and maiming large mammals, so with predatory lizards up to 3 metres (10 feet) long and weighing more than 70 kilograms (150 pounds), these are not good places to lie around with mobility issues!

My first view of wild Komodo Dragons was like a vision from hell. A trek up a dry creek bed in the hills of Rinca Island led to a stagnant pool containing a dead buffalo, with a horn and rib cage protruding from water black with putrefaction. Beside the pool lay an immense Komodo Dragon. A sharp dark line along its flanks indicated the high-water mark where it had been wading in to rip bits off the carcass.

Over the next hour several more dragons arrived to taste the delight. Their deeply forked, constantly protruding tongues are so sensitive to olfactory cues that a rotting buffalo can lure them in from up to 3 kilometres (2 miles) away.

Komodo Dragons are ambush predators, often launching attacks from beside well-used animal trails to inflict ripping bites. They then follow their prey until it dies or is too debilitated to defend itself. Several Komodo Dragons often gather to rip the animal apart.

Adult Komodo Dragons prey on buffalo, deer and pigs, all introduced by humans. An extinct pygmy elephant may once have played a role in their ecology. Fortunately attacks on humans are exceedingly rare.

There has been debate about just how Komodo Dragons subdue their prey. Venom or bacteria-laden saliva? Recent research has revealed venom that lowers blood pressure, causes massive bleeding, prevents clotting and induces shock. But there are also pendulous globules of saliva hanging from the lizards' mouths. Having watched them rip a rotten buffalo apart, I need convincing that the dangling drool is not an appalling cocktail of dangerous microbes.

Komodo Dragons were once hailed as an example of island gigantism, a trend where extremely large sizes evolve in insular environments. However, fossil records now clearly place Komodo Dragons in Australia, along with an even larger monitor called Megalania (*Varanus priscus*). Indonesia's Lesser Sunda Islands may be the last bastions, where a formerly extensive distribution has contracted to just a few islands.

Today Komodo Dragons are Endangered. Being drawcards for tourists is a double-edged sword. There is one financial imperative to conserve them, and another to draw more tourists and increase infrastructure. And with more people come unwanted visitors. If the Asiatic Common Toad (*Duttaphrynus melanostictus*) arrives and becomes established, a fast-breeding toxic prey among these ancient predators could be disastrous.

95

ASIAN WATER MONITOR

Varanus salvator

FAMILY:
Varanidae

LOCATION:
Sungei Buloh Wetland Reserve, Singapore

STATUS:
Least Concern, IUCN Red List

It is often surprising to see large monitor lizards thriving in densely populated human environments. Asian Water Monitors are common around temples in Sri Lanka, in wetland reserves and urban drainages in Singapore, on beaches in Borneo and around ornamental ponds in botanic gardens in Indonesia. With heads protruding, bodies submerged and limbs tucked to their sides, they cruise the waterways of South-East Asia using slow, easy undulating movements of their long, laterally compressed tails.

Traditionally inhabitants of mangrove swamps, riverbanks and other waterside habitats, these semi-aquatic and arboreal lizards, which grow to more than 2 metres (6 feet), are now quite at home around towns and cities. And they appear none too fussy about water quality, seeming to thrive around water in drainage lines I'd be wary of dipping a toe in.

Crossing water barriers is no problem. Their vast distribution across South-East Asia includes many islands. They were among the first colonists of the Krakatau Islands in Indonesia, which were blasted into existence following a huge volcanic eruption in the 1800s. Their early arrival within 25 years of the eruption and their present abundance on all islands in the group is testament to their success as versatile colonisers and their ability to make sea crossings.

These generalised carnivores have an extremely varied diet, largely influenced by where they live. I watched them catch and eat crabs among the mangroves of Singapore's Sungei Buloh Wetland Reserve, while on some beaches they are common predators of turtle eggs and hatchlings. Efforts to conserve dwindling turtle numbers sometimes involve excluding monitor lizards from the nest areas.

Water monitors living in the direct vicinity of human settlements largely subsist on organic waste among the garbage. The fortunes of those near resorts in Malaysia rise and fall with the tourist season when abundant discarded scraps are available. Those lizards living the easy life have a tendency to become obese.

When giant predatory lizards occur in close contact with humans there are bound to be conflicts. They are notorious raiders of poultry, taking both the birds and eggs. In turn, over much of their range Asian Water Monitors are captured for their meat.

While the species is regarded as 'Least Concern' under IUCN listing, there is heavy exploitation of them for their hides. The Indonesian leather industry exports around 40,000 skins per year, which could have serious impacts at population levels.

The Asian Water Monitor successfully takes advantage of human disturbance and it reaps the benefits of society's discarded nutrients. At the same time it endures protracted pressure for its meat and skin. For this lizard, life is a double-edged sword.

Reference: 67

96

RUSTY DESERT MONITOR

Varanus eremius

FAMILY:

Varanidae

LOCATION:

Barkly Tableland,
Northern Territory, Australia

STATUS:

Least Concern, IUCN Red List

The tracks etched in the red desert sand were conspicuous and distinctive. A continuous straight tail drag-mark edged by regularly spaced clawed footprints.

Following the tracks, it was clear the lizard was hunting, taking a straight course across an open expanse and moving closely around the margins of the spinifex hummocks. Then those tracks did something extraordinary. They veered towards another track, apparently to investigate it, then overlayed it by moving across it. I had only been on site for about 10 minutes but that other track was one of my own footprints!

That was my first experience with the common and widespread but extremely elusive Rusty Desert Monitor. Its range extends across the immense spinifex deserts of Australia's west and interior, but as testament to its secretive nature it is much more common to encounter its tracks rather than a lizard in the flesh.

Studies have shown the Rusty Desert Monitor forages so widely that if there is a home range it is extremely large. It investigates burrows and other soil disturbances, including any human diggings, and appears to have a precise map of the terrain including burrow locations. If pursued, it can dash directly to the closest one.

It feeds mainly on lizards, and in the Australian deserts there is a great variety to pick from. Nearly all recorded prey species are diurnal, suggesting it primarily hunts by pursuing and ambushing active lizards rather than extracting animals from their shelter sites. It is also fond of large insects such as grasshoppers.

For many years this species was my Holy Grail of the arid zone – a mysterious, almost mythical creature that lurked on the periphery of my vision. It left traces but never revealed itself. Even now, looking back on the few occasions I have seen it active, the Rusty Desert Monitor is usually a red blur streaking at high speed from a spinifex clump to vanish without trace.

My next mission, should I decide to accept, is to observe one as it forages and hunts undisturbed on the red sand among the spinifex. To see it actually making tracks without fleeing! That is a tough assignment, but not impossible.

Reference: 58

97

KIMBERLEY ROCK MONITOR

Varanus glauerti

FAMILY:

Varanidae

LOCATION:

El Questro, Western Australia

STATUS:

Least Concern, IUCN Red List

Sandstone cliffs and gorges of the Kimberley region of north-western Australia are deeply fissured with vertical and horizontal crevices, studded with caves, ledges and overhangs, and capped with rock pavements. Assuming such landforms play a significant role in shaping some of the animals that inhabit them, the Kimberley Rock Monitor is one of the great masterpieces.

To see a Kimberley Rock Monitor functioning in that stunning and challenging landscape is to behold the magnificence of structural perfection. With a narrow flat head and long neck, depressed body, long limbs and a thin, whip-like tail, this gravity-defying acrobat is superbly designed to travel easily over all dimensions of its complex rocky environment. Foraging monitors are hyperalert, swift and agile as they scale vertical faces, explore narrow crevices, and execute well-coordinated leaps across gaps between rocks.

It has been suggested that the Kimberley Rock Monitor finds most of its prey by searching concealment sites. It is an active investigator of nooks, crannies and vegetation adjacent to the rocks. It appears to feed mainly on invertebrates such as grasshoppers and spiders, and also small lizards.

I have only ever seen a few Kimberley Rock Monitors in the wild, usually curled deep within crevices or moving quickly out of reach among boulders. However, while undertaking fieldwork in the Kimberleys, I became aware of one that used to bask and forage regularly along a rockface that could be accessed by a narrow ledge in a small gorge. Repeated attempts to sneak up with my camera invariably failed for the same simple reason. It could always see me coming.

Success was eventually achieved by embarking on a lengthy detour, climbing the gorge some distance from the basking lizard and stealthily approaching it from above, where there were boulders and vegetation to mask my approach.

I knew any noise or sudden movement would probably terminate the photo session immediately, but sometimes a mistake can have a fortuitous outcome. The lizard was on a ledge above the water, lying flat, and turned slightly away when I accidentally trod on a stick. Instantly the head was raised high on the long slender neck, and cocked towards me as it sensed my approach. That tense frozen moment allowed me a few precious seconds to take the pictures I wanted. Then the show was over.

Reference: 83

98

MERTENS' WATER MONITOR

Varanus mertensi

FAMILY:

Varanidae

LOCATION:

Howard Springs,
Northern Territory, Australia

STATUS:

Endangered, IUCN Red List

Dense pandanus borders a still waterway under a canopy of tall paperbarks. The humid air feels thick and syrupy, and just the faintest breeze rustles the tough serrated foliage. Stalking along the riverbank, it is difficult to tread quietly on a carpet of crunchy dead pandanus leaves. Suddenly a noisy explosion of leaf litter, the scrambling of claws on bark and a loud plop. A startled Mertens' Water Monitor has dashed up an overhanging branch and dropped into the water.

This has always been the familiar scenario along the tropical watercourses of northern Australia, which are home to this semi-aquatic and arboreal monitor. With its laterally compressed tail, nostrils placed high on the snout, muscular body and strong claws, it is a powerful swimmer and an expert climber.

Water monitors hunt in and around the water for anything they can catch or scavenge. This includes birds' eggs and fledglings, discarded fish scraps, frogs and freshwater crabs. They even walk under water, exploring objects and cavities with their tongues in the same way they do on land, and can locate and excavate the submerged nests of long-necked turtles.

But this has all changed. Monitor numbers have crashed dramatically, and the species has completely disappeared from many areas where it was once considered ubiquitous. This sad scenario is likely to deteriorate.

The culprit is the invasive and highly toxic Cane Toad (*Rhinella marina*), which was introduced in the Queensland cane fields in 1935 to control the Cane Beetle (an idea which proved to be a dismal failure). No studies of the toad's potential impact on the environment were carried out, and one cannot conceive a more disastrous and futile exercise in bio-control. Toads have bred prolifically and rapidly spread across northern Australia. They now number in the billions and attempts to prey on them usually have fatal consequences.

Mertens' Water Monitor is one of the most hard-hit species. Almost the entirety of its distribution lies within the predicted range of the Cane Toad. A decline of up to 80 per cent of the monitor population is expected in coming years. A cascading consequence has been the disruption of established predator-prey relationships. For example the Crimson Finch (*Neochmia phaeton*) has been a winner. With fewer monitors to raid nests there has been a spike in fledgling success.

The great rivers and lagoons of northern Australia still look the same, but that scrambling in the leaf litter, scratching on bark and plopping into the water has gone quiet. Healthy monitor populations do remain in some areas, despite the amphibian onslaught. Those fragile strongholds are priceless and deserve whatever protection we can offer.

References: 17, 24, 95

99

NORTHERN RIDGE-TAILED MONITOR

Varanus primordius

FAMILY:

Varanidae

LOCATION:

Adelaide River,
Northern Territory, Australia

STATUS:

Least Concern, IUCN Red List

Monitors are the world's largest lizards. Giant predators and scavengers roam the lands and waterways from Africa and the Middle East, through Asia to Oceania. As the stronghold of the family, Australia has its share of enormous lizards over most parts of the continent.

But some Australian monitors have evolved along a quite different course. They became small. More than two thirds of Australia's monitors do not exceed 1 metre (3 feet) and some reach total lengths of only 40 centimetres (16 inches) or less. Like miniature replicas they have the same coarse loose skin, deeply forked tongues for chemoreception and sharp, recurved teeth. While adults of large species have essentially outgrown most predators, that is not the case for these secretive 'nano-monitors', so the swaggering gait of their giant cousins is replaced by a more furtive dash. Instead of feeding on carrion and large prey they eat insects and small lizards.

Pygmy monitors include arboreal species that hide under bark and in hollows, desert lizards that burrow in sand under spinifex and rock-inhabitants that wedge themselves into crevices and burrow under stones. With a total length of just 25 centimetres (10 inches), the Northern Ridge-tailed Monitor is one of the smallest.

These monitors live in the tropical savannah woodlands and rock outcrops in the north-western Northern Territory. The thick tail, armoured with tough, strongly keeled scales, is consistent with that of several other small to medium-sized monitors associated with rocks or cavities in hard soils. Northern Ridge-tailed Monitors dig shallow burrows under stones and other surface debris, and the tail is used to block the passage to render the lizard difficult to extract.

Although common but patchy within a restricted distribution, Northern Ridge-tailed Monitors are extremely furtive and rarely witnessed foraging. Their highly secretive behaviour mirrors that of other pygmy monitors and the juveniles of large species before they have grown to a 'safer size'. Tiny monitors are almost never seen on the move, except for an occasional fleeting glimpse, even by the most astute herpetologists. Most Northern Ridge-tailed Monitors are encountered during active searches when objects are lifted.

The forces driving miniaturisation in Australian monitors may lie with mammals, or rather the absence of small to medium-sized diurnal predatory species. There is no shortage of these across the global range of monitors, but in Australia mammalian predators, from quoll to planigale, are nocturnal. It could be that nature abhors a vacuum so monitors, widely attributed with a mammal-like intelligence, shrunk and stepped in to fill the gap.

Reference: 38

LACE MONITOR

Varanus varius

FAMILY:

Varanidae

LOCATION:

Great Sandy National Park,
Queensland, Australia

STATUS:

Least Concern, IUCN Red List

Lace Monitors are a familiar sight in forested parts of eastern Australia. They are particularly fond of picnic grounds where, with flaking hides and swaggering gaits, the huge lizards roam between picnic tables and barbecues looking for scraps.

However the juveniles are another matter. They are rarely seen. I can just about count on one hand my sightings of very small individuals. The young must be reasonably common but they are extremely secretive. It seems they live a furtive existence in the trees until, at around 75 centimetres (30 inches) in total length, they become bold enough to show themselves.

As luck would have it, I once stumbled across a juvenile wrestling with a giant centipede in the middle of a walking track. The lizard certainly knew the business end of its prey, for it had a secure grip on the centipede's head. The centipede was nearly as long as the monitor's torso, and with plenty of gripping legs it was not easy to subdue. The ensuing battle lasted half an hour.

Together, reptile and invertebrate rolled in the leaf litter, each with a firm grip on the other. The lizard's tactic was to maintain hold of the head, complete with large fangs and powerful venom, and wipe the centipede across the ground to disengage the grip of its multiple legs. The centipede's option was to hang on, at one stage gripping the full length of the monitor's throat. The whole thing ended surprisingly abruptly when the centipede came unstuck. That allowed the monitor a few jerks of the head and swipes on the ground to gain a better hold, before quickly swallowing it.

This remains one of my treasured herpetological moments, made all the more special as I was able to photograph the whole sequence. The lizard had been so engaged in overcoming its formidable prey that it had been completely indifferent to my presence. When I left the young Lace Monitor it had found a patch of sun and it was basking, alert with head high and a full belly.

REFERENCES

1. Adamopoulou, C., Valakos, E. and Pafilis, P., 1999. Summer diet of *Podarcis milensis*, *P. gaigeae* and *P. erhardii* (Sauria: Lacertidae). *Bonn Zoological Bulletin* **48** (3-4): 275–282.

2. Alexander, J. E., 1838. *An expedition of discovery into the interior of Africa through the hitherto undescribed countries of the Great Namaquas, Boschmans and Hill Damaras.* **1–2**. Henry Colburn Publisher, London.

3. Andrews, R., 1983. *Norops polylepis* (largartija, anole, anolis lizard). Pp. 409–410 in Janzen, D. (ed.). *Costa Rican Natural History.* University of Chicago Press.

4. Armstead, J., Ayala-Varela, F., Torres-Caravajal, O., Ryan, M. and Poe, S., 2017. Systematics and ecology of *Anolis biporcatus* (Squamata: Iguanidae). *Salamandra* **53** (2): 285–293.

5. Austin, C. and Jessing, K., 1994. Green-blood pigmentation in lizards. *Comparative Biochemistry and Physiology* **109** (3): 619–626.

6. Baier, F., Sparrow, D. and Wiedl, H., 2013. *The Amphibians and Reptiles of Cyprus.* Edition Chimaira.

7. Bambaradeniya, C., Samarawickrema, P. and Ranawana, K., 1997. Some observations on the natural history of *Lyriocephalus scutatus* (Linnaeus, 1776) (Reptilia: Agamidae). *Lyriocephalus,* **3** (1): 25–28.

8. Bartlett, R. and Bartlett, P., 2003. *Reptiles and Amphibians of the Amazon – an Ecotourist's Guide.* University Press of Florida.

9. Bauer, A., 2013. *Geckos. The Animal Answer Guide.* Johns Hopkins University Press.

10. Bauer, A., Jackman, T., Sadlier, R. and Whitaker, A., 2009. Review and phylogeny of the New Caledonian diplodactylid gekkotan genus *Eurydactylodes* Wermuth, 1965, with the description of a new species. In Grandcolas, P. (ed.). Zoologia Neocaledonica 7. Biodiversity studies in New Caledonia. *Mem. du Museum National d'Histoire Naturelle* **198**: 13–36.

11. Bauer, A. and Sadlier, R. 2000. *The Herpetofauna of New Caledonia.* Society for studying amphibians and reptiles.

12. Beck, D., 2009. *Biology of Gila Monsters and Beaded Lizards.* University of California Press.

13. Benavides, E., Baum, R., Snel, H. and Sites, J., 2009. Island biogeography of Galápagos lava lizards (Tropiduridae: *Microlophus*): species diversity and colonization of the archipelago. *Evolution* **63** (6): 1606–1626, doi: 10.1111/j.1558-5646.2009.00617.x

14. Boulenger, G., 1888. Descriptions of new reptiles and batrachians from Madagascar. *Ann. Mag. Nat. Hist.* 6 (**1**): 101–107.

15. Branch, B., 1998. *Field Guide to Snakes and Other Reptiles of Southern Africa.* Third Edition. Struik Publishers.

16. Chapple, D., 2003. Ecology, life-history, and behavior in the Australian scincid genus *Egernia*, with comments on the evolution of complex sociality in lizards. *Herpetological Monographs,* **17**, 2003, 145–180.

17. Chapple, D., Tingley, R., Mitchell, N., Macdonald, S., Keogh, J., Shea, G., Bowles, P., Cox, N. and Woinarski, J., 2019. *The Action Plan for Australian Lizards and Snakes 2017.* CSIRO Publishing.

18. Cuellar, O. 1976. Intraclonal histocompatibility in a parthenogenetic lizard: evidence of genetic homogeneity. *Science* **193** (4248): 150–153.

19. Das, I., 1995. 'Aspects of the biodiversity and biogeography of Sri Lanka'. *Lyriocephalus,* **1** (1-2): 17–20.

20. Das, I., 2010. *A Field Guide to Reptiles of South-East Asia.* New Holland Publishers.

21. Das, I., Charles, J. and Edwards, D., 2008. *Calotes versicolor* (Squamata: Agamidae) – A new invasive squamate for Borneo. *Current Herpetology* **27** (2): 109–112.

22. D'Cruze, N., Köhler , J., Franzen, M. and Glaw, F. 2008. A conservation assessment of the amphibians and reptiles of the Forêt d'Ambre Special Reserve, north Madagascar. *Madagascar Conservation & Development.* **3**: (1): 44–54.

23. De Silva, A., Bauer, A., Austin, C., Goonewardene, S., Hawke, Z., Vanneck, V., Drion, A., de Silva, P., Perera, B., Jayaratne, R. and Goonasekera, M., 2004. 'Distribution and natural history of *Calodactylodes illingworthorum* (Reptilia: Gekkonidae) in Sri Lanka: Preliminary findings.' Pp. 192–198. In De Silva, A. (ed). The herpetology of Sri Lanka: Current research. *J. of the Amphibia and Reptile Research Organization of Sri Lanka (ARROS).* **5**: 1–2.

24. Doody, J. S., Sloans, R., Castellano, C., Rhind, D., Green, B., McHenry, C. and Clulow, S., 2015. Invasive toads shift predator–prey densities in animal communities by removing top predators. *Ecology* **96**, 2544–2554.

25. Doughty, P., Kealley, L., Shoo, L. and Melville, J., 2015. Revision of the Western Australian pebble-mimic dragon species-group (*Tympanocryptis cephalus*: Reptilia: Agamidae). *Zootaxa* 4039 (**1**): 85–117.

26. Duellman, W., 1978. The biology of an equatorial herpetofauna in Amazonian Ecuador. *Miscellaneous Publications, Museum of Natural History, University of Kansas.* **65**: 1–352

27. Eckhardt, E., Kappeler, P. and Kraus, M., 2017. Highly variable lifespan in an annual reptile, Labord's chameleon (*Furcifer labordi*). *Scientific Reports* 7, 11397.

28. Eifler, D. and Eifler, M. 2014. Escape tactics in the lizard *Meroles cuneirostris. Amphibia Reptilia* **35**: 393–389.

29. Ehmann, H., 1982. The natural history and conservation status of the Adelaide Pigmy Bluetongue Lizard *Tiliqua adelaidensis. Herpetofauna* **14** (1): 61–76.

30. Greer, A., 1989. *The Biology and Evolution of Australian Lizards.* Surrey Beatty and Sons.

31. Greer, A. and Raizes, G., 1969. Green blood pigment in lizards. *Science* **166** (3903): 392–393.

32. Grismer, L., 2002. *Amphibians and Reptiles of Baja California including its Pacific Islands in the Sea of Cortes.* University of California Press.

33. Grismer, L., 2011. *Lizards of Peninsular Malaysia, Singapore and their adjacent archipelagos.* Edition Chimaira.

34. Hagen, I. and Bull, M., 2011. Home ranges in the trees: Radiotelemetry of the Prehensile Tailed Skink, *Corucia zebrata. Journal of Herpetology* 45 (**1**): 36–39.

35. Hansen, D. and Müller, C., 2009 a. Reproductive ecology of the endangered enigmatic Mauritian endemic *Roussea simplex* (Roussaceae). *International Journal of Plant Sciences* 170(1): 42–52.

36. Hansen, D. and Müller, C., 2009 b. Invasive ants disrupt gecko pollination and seed dispersal of the endangered plant *Roussea simplex* in Mauritius. *Biotropica* **41** (2): 202–208

37. Herrel, A., Huyghe, K., Vanhooydonck, B., Backeljau, T., Breugelmans, K. Grbac , I., Van Damme, R. and Irschick, D., 2008. Rapid large-scale evolutionary divergence in morphology and performance associated with exploitation of a different dietary resource. *Proceedings of the National Academy of Sciences of the United States of America.* **105** (12): 4792–5.

38. Husband, H. and Christian, K., 2004. *Varanus primordius.* Pp 434–437 In Pianka, E., King., D. and King., R. (eds.). *Varanoid Lizards of the World.* Indiana University Press.

39. Hutchinson, M., Adams, M. and Ehmann, H., 2007. *Aprasia aurita* (Squamata, Pygopodidae), an addition to the lizard fauna of South Australia. *Herpetofauna* **37** (2): 98–103.

40. Hutchinson, M., Couper, P., Amey, A. and Worthington Wilmer, J., 2021. Diversity and systematics of limbless skinks (*Anomalopus*) from eastern Australia and the skeletal changes that accompany the substrate swimming body form. *Journal of Herpetology* **55** (4): 361–384.

41. Ito, R. and Mori, A., 2012. The Madagascan spiny-tailed iguana alters the sequence of anti-predator responses depending on predator types. *African J. Herpet.* **61** (1): 58–68.

42. Jayasekara, E., Mahaulpatha, W. and De Silva, A., 2018. Habitat utilization of endangered Rhino Horned Lizard (*Ceratophora stoddartii*) (Sauria: Agamidae) in the Horton Plains National Park, Sri Lanka. *Sri Lanka Journal of Entomology and Zoology Studies* **6** (4): 1544–1549.

43. Kornilios, P., Thanou, E., Lymberakis, P., Ilgaz, C., Kumlutas, Y. and Leache, A., 2019. A phylogenomic resolution for the taxonomy of Aegean green lizards. *Zoologica Scripta* **49** (1): 14–27.

44. Leenders, T., 2019. *Reptiles of Costa Rica: A Field Guide.* Cornell University Press.

45. Linkem, C., Brown, R., Siler, D., Evans, B., Austin, C., Iskandar, D., Diesmos, A., Supriatna, J., Andayani, N., and Mcguire, J., 2013. Stochastic faunal exchanges drive diversification in widespread Wallacean and Pacific island lizards (Squamata: Scincidae: *Lamprolepis smaragdina*). *Journal of Biogeography*, 40 (3): 507–520. https://doi.org/10.1111/jbi.12022

46. Littleford-Colquhoun, B., Clemente, C., Whiting, M., Ortiz-Barrientos, D. and Frere, C., 2017. Archipelagos of the anthropocene: rapid and extensive differentiation of native terrestrial vertebrates in a single metropolis. *Molecular Ecology* **26** (9): 2466–2481.

47. Luiselli , L. and Capizzi, D., 1999. Ecological distribution of the geckos *Tarentola mauritanica* and *Hemidactylus turcicus* in the urban area of Rome in relation to age of buildings and condition of the walls. *Journal of Herpetology* **33** (2): 316–319.

48. Matyot, P., 2004.The establishment of the crested tree lizard, *Calotes versicolor* (Daudin, 1802) (Squamata: Agamidae), in Seychelles. *Phelsuma* **12**: 35–47.

49. McCoy, M., 2006. *Reptiles of the Solomon Islands*. Pensoft.

50. McGrath, J., 2008. 'Geckos: Family Gekkonidae.' Pp. 73–213 In Swan, M., (ed). *Keeping and Breeding Australian Lizards*. Mike Swan Herp Books.

51. Melville, J. and Wilson, S., 2019. *Dragon Lizards of Australia: Evolution, Ecology and a Comprehensive Field Guide*. Museums Victoria Publishing.

52. Mitchell, F., 1973. Studies on the ecology of the agamid lizard *Amphibolurus maculosus* (Mitchell). *Trans. R. Soc. S. Aust.* **97** (1): 47–76.

53. Mouton, P. Le F., Flemming, A. and Kanga, M., 2006. Grouping behaviour, tail-biting behaviour and sexual dimorphism in the armadillo lizard (*Cordylus cataphractus*) from South Africa. *J. Zool.* **249** (1): 1–10.

54. Murray, G. and Schramm, D., 1987. A comparative study of the diet of the wedge-snouted sand lizard *Meroles cuneirostris* (Strauch) and the sand diving lizard, *Aporosaura anchietae* (Bocage) (Lacertidae), in the Namib Desert. *Madoqua* **15** (1): 55–61.

55. Noonan, B, and Sites J., 2010. Tracing the origins of iguanid lizards and boine snakes of the Pacific. *The American Naturalist*: **175** (1): 61–72.

56. Oliver, P., Travers, S., Richmond, J., Pikacha, P. and Fisher, R., 2017. At the end of the line: independent overwater colonizations of the Solomon Islands by a hyperdiverse trans-Wallacean lizard lineage (*Cyrtodactylus*: Gekkota: Squamata). *Zoological Journal of the Linnean Society*, **182** (3): 681 694. https://doi.org/10.1093/zoolinnean/zlx047

57. Patchell, F. and Shine, R., 1986. Feeding mechanisms in pygopodid lizards: How can *Lialis* swallow such large prey? *J. Herpet.* **20** (1): 59–64.

58. Pianka, E., 1968. Notes on the biology of *Varanus eremius*. *West Australian Naturalist* **11** (2): 39–44.

59. Pianka, E., 1969. Sympatry of desert lizards (*Ctenotus*) in Western Australia. *Ecology* **50**: 1012–1030.

60. Pianka, E., 1970. Comparative autecology of the lizard *Cnemidophorus tigris* in different parts of its geographic range. *Ecology* **51** (4): 703–720.

61. Pianka, E. and Pianka, H., 1976. Comparative ecology of twelve species of nocturnal lizards (Gekkonidae) in the Western Australian desert. *Copeia,* 1976 (1): 125–142.

62. Pianka, E. and Vitt, L., 2003. *Lizards – Windows to the Evolution of Diversity*. University of California Press.

63. Pough, F. H. (1973). Lizard energetics and diet. *Ecology* **54** (4): 837–844.

64. Prötzel, D., Martin Hess, M., Schwager, M., Glaw, F. and Scherz, M., 2021. Neongreen fluorescence in the desert gecko *Pachydactylus rangei* caused by iridophores. *Sci Rep* **11**: 297. https://doi.org/10.1038/s41598-020-79706-z

65. Pui, Y. and Das, I., 2012. A significant range extension for the Kinabalu Parachute Gecko, *Ptychozoon rhacophorus* (Boulenger, 1899) (Squamata: Gekkonidae) and a new state record from Sarawak, north-western Borneo. *Herpetology Notes* **6**: 177–179.

66. Ramanantsoa, G., 1984. 'The Malagasy and the chameleon: A traditional view of nature.' Pp. 205–209 In Jolly, A., Oberle, P. and Albignac, R. (eds). *Key Environments, Madagascar*. Pergamon Press.

67. Rawlinson, P., Widjoya, A., Hutchinson, M. and Brown, G.,1990. The terrestrial vertebrate fauna of the Krakatau Islands, Sunda Strait, 1883-1986. *Philosophical Transactions of the Royal Society London B.* **328**: 3–28.

68. Robinson, B and Barrows, C., 2013. Namibian and North American sand-diving lizards. *Journal of Arid Environments* **93**:116–125.

69. Robinson, M. and Cunningham, A. 1978. Comparative diet of two Namib Desert sand lizards (Lacertidae). *Madoqua* **11**: 41–53.

70. Russell, A. and Bauer, A., 1990. Substrate excavation in the Namibian web-footed gecko, *Palmatogecko rangei* Andersson 1908, and its ecological significance. *Tropical Zoology* **3**: 197–207.

71. Savage, J., 2002. *The Amphibians and Reptiles of Costa Rica – A herpetofauna between Two Continents between Two Seas*. University of Chicago Press.

72. Schmidt, J., 2019. Predator–prey battle of ecological icons: horned lizards (*Phrynosoma* spp.) and harvester ants (*Pogonomyrmex* spp.). *Copeia* **107** (3): 404–410.

73. Schutz, L., Stuart-Fox, D. and Whiting, M., 2007. Does the lizard *Platysaurus broadleyi* aggregate because of social factors? *J. Herpet.* **41** (3): 354–359.

74. Seipp, R. and Henkel, F., 2000. Rhacodactylus: *Biology, Natural History and husbandry*. Edition Chimaira.

75. Sherbrooke, W., 2003. *Introduction to Horned Lizards of North America*. University of California Press.

76. Shuttleworth, C., Mouton, P. le F., and van Wyk, H., 2008. Group size and termite consumption in the armadillo lizard, *Cordylus cataphractus*. *Amphibia-Reptilia* **29** (2): 171–176.

77. Snyder, J., Snyder, L., and Bauer, A., 2010. Ecological observations on the Gargoyle Gecko, *Rhacodactylus auriculatus* (Bavay, 1869), in southern New Caledonia. *Salamandra* **46** (1): 37–47.

78. Somaweera, R., and Somaweera, N., 2009. *Lizards of Sri Lanka: A Colour Guide with Field Keys*. Edition Chimaira.

79. Spawles, S., Howell, K., Drewes, R. and Ashe, J., 2004. *A Field Guide to Reptiles of East Africa*. A&C Black.

80. Stapley, J. and Whiting, M., 2006. Ultraviolet signals fighting ability in a lizard. *Biol. Lett.* **2** (2): 169–172

81. Stebbins, R., 2003. *A Field Guide to Western Reptiles and Amphibians*. Third edition. Peterson Field Guides.

82. Stuart-Fox, D., Moussalli, A., Marshall, N. and Owens, I., 2003. Conspicuous males suffer higher predation risk: visual modelling and experimental evidence from lizards. *Animal Behaviour* **66** (3): 541–550.

83. Sweet, S. 2004. *Varanus glauerti*. Pp 366–372 In Pianka, E., King, D. and King., R. (eds). *Varanoid Lizards of the World*. Indiana University Press.

84. Tapia, W. and Gibbs, J., 2022. Galápagos land iguanas as ecosystem engineers. PeerJ 2022 Jan 20;10:e12711. doi: 10.7717/peerj.12711. eCollection 2022.

85. Tilbury, C., 2010. *Chameleons of Africa. An Atlas including the Chameleons of Europe, Middle East and Asia*. Edition Chimaira.

86. Valakos, E., Pafilis, P., Sotiropoulos, K., Lymberakis, P., Maragou, P. and Foufopoulos, J., 2008. *The Amphibians and Reptiles of Greece*. Edition Chimaira.

87. Vanderduys, E., 2017. A new species of gecko (Squamata: Diplodactylidae: *Strophurus*) from central Queensland, Australia. *Zootaxa* **4347** (2): 316–330.

88. van Winkel, D., Baling, M. and Hitchmough, R., 2018. *Reptiles and Amphibians of New Zealand: A Field Guide*. Auckland University Press.

89. Vitt, L., 1991. Ecology and life history of the scansorial arboreal lizard *Plica plica* (Iguanidae) in Amazonian Brazil. *Canadian J. Zool.* **69** (2): 504–511 https://doi.org/10.1139/z91-077

90. Wareham, D., 2007. *The Reptiles and Amphibians of Dorset*. British Herpetological Society.

91. Webb, J. and Shine, R., 1994. Feeding habits and reproductive biology of Australian pygopodid lizards of the genus *Aprasia*. *Copeia* 1994 (**2**): 390–398.

92. Welt, R. and Raxworthy, C., 2022. Dispersal, not vicariance, explains the biogeographic origin of iguanas on Madagascar. *Molecular Phylogenetics and Evolution*. **167** February 2022: 107345.

93. Werner, Y., 2016. *Reptile Life in the Land of Israel*. Edition Chimaira.

94. Wikelski, M. and Thom, C., 2000. Marine iguanas shrink to survive El Niño. *Nature* **403**: 37–38 (2000). https://doi.org/10.1038/47396

95. Wilson, S., 2012. *Australian Lizards: A Natural History*. CSIRO Publishing.

96. Wilson, S. and Knowles, D., 1988. *Australia's Reptiles: A Photographic Reference to the Terrestrial Reptiles of Australia*. HarperCollins.

97. Wilson, S. and Swan, G., 2021. *A Complete Guide to Reptiles of Australia*. Sixth Edition. New Holland Publishers.

98. Zug, G., 2013. *Reptiles and Amphibians of the Pacific Islands*. University of California Press.

INDEX

Published in 2024 by Reed New Holland Publishers
Sydney

Level 1, 178 Fox Valley Road, Wahroonga, NSW 2076, Australia

newhollandpublishers.com

A record of this book is held at the National Library of Australia.

ISBN 978 1 92554 698 9

Managing Director: Fiona Schultz
Publisher and Project Editor: Simon Papps
Designer: Andrew Davies
Production Director: Arlene Gippert

Printed in China

10 9 8 7 6 5 4 3 2 1

Keep up with Reed New Holland
and New Holland Publishers

[f] ReedNewHolland
[Instagram] @NewHollandPublishers and @ReedNewHolland

Front cover. Flap-necked Chameleon, *Chamaeleo dilepis*.
Etosha National Park, Namibia

Back cover. Eastern Water Dragon, *Intellagama lesueurii lesueurii*.
Brisbane, Queensland, Australia

Pages 2–3. Gibber Dragon. *Ctenophorus gibba*.
Maree, South Australia

Page 4. Shrubland Pale-flecked Morethia. *Morethia obscura*.
Perth, Western Australia

Page 6. Eastern Water Dragon, *Intellagama lesueurii lesueurii*.
Brisbane, Queensland, Australia

Page 7. Augrabies Flat Lizard. *Platysaurus broadleyi*.
Augrabies Falls National Park, South Africa

OTHER NATURAL HISTORY TITLES BY REED NEW HOLLAND INCLUDE:

A Complete Guide to Reptiles of Australia Sixth Edition
Steve Wilson and Gerry Swan
ISBN 978 1 92554 671 2

A Field Guide to Reptiles of New South Wales Fourth Edition
Gerry Swan
ISBN 978 1 92554 608 8

A Field Guide to Reptiles of Queensland Third Edition
Steve Wilson
ISBN 978 1 92151 748 8

A Tribute to the Reptiles and Amphibians of Australia and New Zealand
Australian Herpetological Society
Edited by Chris Williams and Chelsea Maier
ISBN 978 1 92554 659 0

Crocodiles of the World
Colin Stevenson
ISBN 978 1 92554 681 1

Reed Concise Guide: Lizards of Australia
Steve Wilson
ISBN 978 1 92554 657 6

Reed Concise Guide: Snakes of Australia
Gerry Swan
ISBN 978 1 92151 789 1

Reptiles and Amphibians of Australia, New Zealand and New Guinea
Australian Herpetological Society
Edited by Chris Williams and Chelsea Maier
ISBN 978 1 92554 673 6

Tadpoles and Frogs of Australia Second Edition
Marion Anstis
ISBN 978 1 92554 601 9

For details of these books and hundreds of other Natural History titles see
newhollandpublishers.com
and follow ReedNewHolland and NewHollandPublishers on Facebook